The Philosophy of Science Fiction Film

The Philosophy of Popular Culture

The books published in the Philosophy of Popular Culture series will illuminate and explore philosophical themes and ideas that occur in popular culture. The goal of this series is to demonstrate how philosophical inquiry has been reinvigorated by increased scholarly interest in the intersection of popular culture and philosophy, as well as to explore through philosophical analysis beloved modes of entertainment, such as movies, TV shows, and music. Philosophical concepts will be made accessible to the general reader through examples in popular culture. This series seeks to publish both established and emerging scholars who will engage a major area of popular culture for philosophical interpretation and examine the philosophical underpinnings of its themes. Eschewing ephemeral trends of philosophical and cultural theory, authors will establish and elaborate on connections between traditional philosophical ideas from important thinkers and the ever-expanding world of popular culture.

Series Editor

Mark T. Conard, Marymount Manhattan College, NY

Books in the Series

The Philosophy of Stanley Kubrick, edited by Jerold J. Abrams
The Philosophy of Film Noir, edited by Mark T. Conard
The Philosophy of Martin Scorsese, edited by Mark T. Conard
The Philosophy of Neo-Noir, edited by Mark T. Conard
The Philosophy of The X-Files, edited by Dean A. Kowalski
The Philosophy of TV Noir, edited by Steven M. Sanders and Aeon J. Skoble
Basketball and Philosophy, edited by Jerry L. Walls and Gregory Bassham

THE PHILOSOPHY OF
SCIENCE
FICTION
FILM

Edited by Steven M. Sanders

THE UNIVERSITY PRESS OF KENTUCKY

Publication of this volume was made possible in part by a grant
from the National Endowment for the Humanities.

Editorial and Sales Offices: The University Press of Kentucky
663 South Limestone Street, Lexington, Kentucky 40508-4008
www.kentuckypress.com

Library of Congress Cataloging-in-Publication Data

The philosophy of science fiction film / edited by Steven M. Sanders.
     p. cm. — (The philosophy of popular culture)
  Includes bibliographical references and index.
  ISBN 978-0-8131-2472-8 (hardcover : alk. paper)
  1. Science fiction films—History and criticism. I. Sanders, Steven, 1945–
  PN1995.9.S26P49 2008
  791.43'615—dc22   2007038230
  ISBN 978-0-8131-9260-4 (pbk. : alk. paper)

CONTENTS

Preface and Acknowledgments   vii

An Introduction to the Philosophy of Science Fiction Film   1
  Steven M. Sanders

Part 1: Enigmas of Identity and Agency

What Is It to Be Human? *Blade Runner* and *Dark City*   21
  Deborah Knight and George McKnight

Recalling the Self: Personal Identity in *Total Recall*   39
  Shai Biderman

Picturing Paranoia: Interpreting *Invasion of the Body Snatchers*   55
  Steven M. Sanders

The Existential *Frankenstein*   73
  Jennifer L. McMahon

Part 2: Extraterrestrial Visitation, Time Travel, and Artificial Intelligence

Technology and Ethics in *The Day the Earth Stood Still*   91
  Aeon J. Skoble

Some Paradoxes of Time Travel in *The Terminator* and *12 Monkeys*   103
  William J. Devlin

2001: A Philosophical Odyssey   119
  Kevin L. Stoehr

Terminator-Fear and the Paradox of Fiction   135
  Jason Holt

Part 3: Brave Newer World: Science Fiction Futurism

The Dialectic of Enlightenment in *Metropolis*   153
  Jerold J. Abrams

Imagining the Future, Contemplating the Past: The Screen Versions of *1984*   171
  R. Barton Palmer

Disenchantment and Rebellion in *Alphaville*   191
  Alan Woolfolk

The Matrix, the Cave, and the Cogito   207
  Mark T. Conard

List of Contributors   223

Index   227

PREFACE AND ACKNOWLEDGMENTS

The essays in this volume explore some of the ideas and possibilities that science fiction films take as their starting points. Since the essays are philosophical, they aim to increase readers' understanding and appreciation by identifying the philosophical implications and assumptions of *The Day the Earth Stood Still, Invasion of the Body Snatchers, The Terminator,* and a dozen other science fiction film classics. The questions these films raise are addressed by philosophers, film theorists, and other scholars who take a variety of approaches and perspectives. No single method or school of thought predominates. Of course, there is a consensus among the contributors that intelligent and well-informed discussion of films can lead to greater appreciation and understanding of them. And each contributor would no doubt agree that it is desirable for readers to have a firsthand acquaintance with the film he or she has chosen to write about.

Aside from being asked to confine their choices to a "short list," described in the introductory essay, contributors were free to treat science fiction films in any way that struck them as illuminating. Some contributors deployed a group of philosophical ideas around their choice of film. Others first selected a philosophical problem or theme, such as time travel, personal identity, or artificial intelligence, and then found a film that was particularly effective at dramatizing and developing the problem or theme in question. Although the essays implicate many areas of philosophy, including ethics, metaphysics, theory of knowledge, political philosophy, and aesthetics, readers who have had no previous exposure to philosophy will almost always be able to pick up the gist of the discussion, if not the finer points of detail. In addition, the introductory essay is designed to clarify the basic line of argument and point of view in each essay. All of the essays involve interpretive "readings" of the films, which means that they invite disagreement and reflection on the basis of that disagreement.

I am fortunate to have worked with colleagues who write about science fiction film so well. I thank them for their patience, hard work, and willingness to share their expertise. I am grateful to Mark T. Conard for developing the series that brings philosophy into such harmonious relationships with popular culture, to Eric Bronson and Michael L. Stephans for their helpful comments during the submission process, and to Christeen Clemens for our discussions of the book from its inception. Finally, I want to thank my editing supervisor, David L. Cobb, and my copyeditor, Anna Laura Bennett, for their valuable suggestions and meticulous correction of the manuscript.

An Introduction to the Philosophy of Science Fiction Film

Steven M. Sanders

Over the last decade there has been a significant shift in the attitudes of philosophers as they have become increasingly receptive to the opportunity to apply methods of philosophical inquiry to film, television, and other areas of popular culture. In fact, *receptive* is far too mild a word to describe the enthusiasm with which many philosophers now embrace popular culture. The authors of the essays included in this volume have genuine affection for science fiction feature films and the expertise to describe, explain, analyze, and evaluate the story lines, conflicts, and philosophically salient themes in them. Their contributions are designed to promote an understanding of the very considerable extent to which philosophy and science fiction are thematically interdependent insofar as science fiction provides materials for philosophical thinking about the logical possibility and paradoxes of time travel, the concept of personal identity and what it means to be human, the nature of consciousness and artificial intelligence, the moral implications of encounters with extraterrestrials, and the transformations of the future that will be brought about by science and technology. Of course, many science fiction films emphasize gadgets and special effects to the neglect of conceptual complexity, but the films discussed here engage viewers on the plane of ideas and provide occasions for historical, political, literary, and cultural commentary as well as philosophical analysis.

This volume includes a dozen philosophically accessible essays on some of the best science fiction films from seven decades. The essays discuss science fiction film classics, and they are classics precisely because they were alive to their own times and are alive to ours as well. In this sense, *Metropolis* (Fritz Lang, 1927), *Frankenstein* (James Whale, 1931), *The Day the Earth*

Stood Still (Robert Wise, 1951), and *Invasion of the Body Snatchers* (Don Siegel, 1956) are acknowledged classics of the genre. The landmark film *2001: A Space Odyssey* (Stanley Kubrick, 1968) continues to influence contemporary filmmakers and awe or baffle viewers forty years after its release. The 1970s, dubbed the decade of "easy riders, raging bulls" by the journalist Peter Biskind in his book of that title, was also the era of the blockbuster science fiction franchise movies *Star Wars* (George Lucas, 1977) and *Star Trek—The Motion Picture* (Robert Wise, 1979). In the 1980s, *Blade Runner* (Ridley Scott, 1982) and *The Terminator* (James Cameron, 1984), science fiction action films with philosophical thrust to spare, were released, and the 1990s had *Total Recall* (Paul Verhoeven, 1990), *Dark City* (Alex Proyas, 1998), and *The Matrix* (Andy Wachowski and Larry Wachowski, 1999), films that remain vital and vibrant.

These films differ significantly in budget, dramatic scope, and imaginative sweep. Most of them were on the editor's short list from which contributors were asked to select a film for discussion. Two criteria guided the choice of films for inclusion on the list. First, the films had to be classics in the sense explained above, and second, they had to be amenable to philosophical examination. Obviously, a case can be made for many films that could not be accommodated within the confines of a single volume, so numerous worthwhile candidates had to be excluded. Naturally, opinions vary on which films should be regarded as science fiction classics, but less so than one might think. On the basis of either box office receipts or critical reception, the place of most of the films discussed in this volume in the science fiction film pantheon seems secure. Their suitability for philosophical interrogation is ably demonstrated by the philosophers, film theorists, and other scholars whose essays constitute case studies in philosophical thinking about popular culture.

Three Types of Philosophical Thinking

The contributors to *The Philosophy of Science Fiction Film* have chosen to address such topics as space, time, causality, consciousness, identity, agency, and other categories of experience. Their essays exhibit three types of philosophical thinking about science fiction films. First, there are essays that develop the historical and intellectual context in which the films were conceived, produced, and received—the latter sometimes by less than comprehending audiences. The cultural understanding and historical erudition that go into

Jerold J. Abrams's essay on *Metropolis,* for example, provide a guide to the constellation of ideas found in the work of the filmmaker Fritz Lang and the philosophers Theodor Adorno and Max Horkheimer. Jennifer L. McMahon develops the literary background and existential themes of *Frankenstein.* And R. Barton Palmer, writing about *1984,* gives us the historical, literary, and philosophical web of thinking that went into both the 1956 and 1984 versions of the film.

Second, there are essays that provide focused analyses of particular films. These essays make explicit the themes, settings, and structure of a specific film and draw out its philosophical implications and assumptions. Aeon J. Skoble's essay on *The Day the Earth Stood Still,* the essay on *Blade Runner* and *Dark City* by Deborah Knight and George McKnight, Mark T. Conard's examination of *The Matrix,* and my own essay on *Invasion of the Body Snatchers* are examples of this type of philosophical thinking about science fiction film.

The third type of philosophical thinking about science fiction film is found in theme-driven essays that use one or more films to motivate philosophical discussion of a particular topic or problem. William J. Devlin's essay uses *The Terminator* and *12 Monkeys* to elucidate two conceptions of time travel. Shai Biderman's essay on *Total Recall* explores alternative conceptions of personal identity. Alan Woolfolk explains disenchantment and rebellion in his essay on *Alphaville,* and Jason Holt discusses how it is possible to be moved to feel genuine emotions about things we know do not exist, the so-called paradox of fiction, in connection with *The Terminator.*

These distinctions among the types of philosophizing provide a framework for understanding the various things the contributors to this volume are doing. While it is useful to distinguish them for theoretical purposes, readers will discover that the three types of philosophical thinking overlap in the work of most of the contributors and are found in each of the essays. Ultimately, the contributors to this volume expose science fiction films to reflection and analysis in order to deepen our understanding of them as well as to introduce readers to the problems, methods, and arguments of philosophy.

In the next section of this introduction, I identify a number of philosophical problems and themes found in the essays and pose critical questions that readers may wish to ask about them. Some readers may find it beneficial to read these comments before reading the individual essays, but others may wish to read the essays first to form their own opinions and then come back to this portion of the introduction to think about my comments. Since I discuss matters that first-time viewers may wish to discover for themselves, let me issue a "spoiler" alert to those who proceed to the next section.

Problems and Themes

The philosophers who write about science fiction films in this volume describe what happens in these films and identify and analyze what is implied. They explain the philosophical arguments, ethical perspectives, and metaphysical ideas that lie behind the images we see on the screen.

Many of the best science fiction films are thought to be allegories and have been interpreted symbolically. For example, *The Day the Earth Stood Still* is called a "slightly veiled story of the life of Christ" by James O'Neill, who parenthetically adds, "I know it's a stretch but it's there if you look for it."[1] In *The Rough Guide to Sci-Fi Movies,* John Scalzi writes, "In the movie aliens send an emissary, named Klaatu, to make contact with us earthlings, and we respond by grievously wounding Klaatu at seemingly every convenient moment. This all points to a blatant Klaatu/Christ analogy, which, incidentally, went right over the head of director Robert Wise, who has professed surprise that people read religious subtexts into the film. And yet the Christ-like qualities are richly in evidence—including Klaatu's idea to go by the name 'Carpenter' while wandering among the humans."[2] Similarly, Kim Newman writes, "Considering screenwriter Edmund H. North's insistent Christ references, we can perhaps assume that the Gorts represent an infallible, divine solution to the nuclear stalemate."[3] However, it is controversial whether, or in what sense, such films treat social, political, or religious issues symbolically. For example, Aeon J. Skoble repudiates the religious interpretation of *The Day the Earth Stood Still.* He argues that there are significant differences between Klaatu and Christ and that, although he does not reject an allegorical interpretation of the film, he rejects this one.

The symbolic character of science fiction films is explained, and in some instances challenged, by other contributors as well. In her essay on *Frankenstein,* Jennifer L. McMahon maintains that it is a primary function of Frankenstein's monster to personify death, with all the ramifications this has for our efforts to prolong life. I point out in my essay on *Invasion of the Body Snatchers* that numerous commentators have claimed that the film is a Cold War allegory of the pervasive red scare of the 1950s.

In part 1, "Enigmas of Identity and Agency," five contributors discuss philosophical questions about the nature of personal identity, moral agency, and what it means to be human. According to Andrew Spicer, in *Blade Runner*, "a hybrid 'future noir' that depicted a nightmare Los Angeles of 2019 as an entropic dystopia characterized by debris, decay, and abandonment," we have a full-blown depiction of the dark and depraved universe of noir science fiction.[4] In "What Is It to Be Human? *Blade Runner* and *Dark City*," Deborah Knight and George McKnight use both films to discuss the role of memory and the emotions as an answer to the question that provides the title to their essay. One of the most influential and controversial science fiction films of the last two and a half decades, *Blade Runner* has been widely imitated and discussed. A chief source of its controversy concerns the fact that the director, Ridley Scott, pulled the theatrical release from the shelves once the film went to video and released an authorized director's cut. It has always been a vexing question whether the blade runner, Rick Deckard (Harrison Ford), is himself a replicant. Knight and McKnight strongly suggest that he is. One of the most convincing pieces of evidence for them is that Deckard appears to have memories implanted in him by the sinister Tyrell Corporation, for which he works. But this would be compelling only if we knew that *no humans* have implanted memories, and it is not clear that we know this. In view of *Blade Runner*'s film noir lineage, it would not be unreasonable to suspect that, as in some classic noir films that feature a protagonist suffering from amnesia, Deckard has had memories implanted in him that he recalls in dreams.[5]

Scott Bukatman, the author of a best-selling study of *Blade Runner*, also weighs in on the question of whether Deckard is human or replicant. Citing *Do Androids Dream of Electric Sheep?*, the novel by Philip K. Dick on which the film *Blade Runner* is based, Bukatman writes, "Deckard confidently locates the difference between humans and their imitators: 'An android doesn't care what happens to another android.' To which someone logically replies, 'Then you must be an android.'"[6] But is this a logical reply? Not if by *logical* one means following the rules of sound reasoning. The facts that (1) an android doesn't care what happens to another android and that (2) Deckard doesn't care what happens to androids do not allow us to conclude that (3) Deckard is, or must be, an android. The error consists in thinking that a feature that applies to androids (they don't care what happens to other androids) and also applies to Deckard (he doesn't care what happens to androids) entails that Deckard is himself an android. Consider the analogous reasoning: (1) All cats are animals and (2) my dog, Spot, is an animal. (3)

Therefore, Spot is a cat. Furthermore, premise (2) of the original argument is false because Deckard does indeed care about Rachael, who he knows is an android. This means that even if Deckard *is* an android, premise (1) is false: some androids *do* care about what happens to another android. Thus the argument is logically invalid, its premises are false, and its conclusion is false.

Knight and McKnight say much in defense of the claim that it is the emotions and desires that prompt action, and it is this, independently of any memories that may have been implanted in Deckard, that explains his change of heart about Rachael, with whom he finds himself falling in love. But is the assumption that we can neatly separate emotions and desires from memories true? Is it even coherent? As Knight and McKnight themselves observe, if one could not remember one's aims, commitments, and values from one moment to the next, action would be all but impossible and relationships could not be sustained.

In connection with *Dark City*, it might be assumed that one can easily identify the body of the protagonist, Murdoch, even if his memories have been tampered with, added to, or stolen. But how does one establish that new memories have been programmed into Murdoch's body without being able to *independently* identify that body as *Murdoch's?* This problem of "other bodies" is less widely discussed by philosophers than the traditional problem of "other minds," but it is just as thought provoking and, arguably, just as relevant to the solution (or dissolution) of that venerable metaphysical problem.[7]

Alternative views of personal identity are discussed by Shai Biderman in "Recalling the Self: Personal Identity in *Total Recall.*" Biderman begins by distinguishing between two questions philosophers raise when they discuss the problem of personal identity. The first is the problem of what constitutes a person at all, that is, an entity of the type *person* as opposed to *nonperson* (like a stone or flower). The second question concerns what makes a person the *same* person over time. In connection with this question, Biderman considers a number of answers in terms of proposed criteria of identity. What makes a person the same person over time, he argues, may be sameness of body, sameness of brain, sameness of memory, or psychological connectedness. Finding difficulties with each of these answers, Biderman turns to the view, prominent among postmodernist philosophers such as Jacques Derrida (though it can be traced to the eighteenth-century philosopher David Hume's notorious "bundle" theory of the self), that the self is fictitious. On this view, the self is not an enduring, substantial entity at all but something socially constructed and therefore capable of being *de*constructed. In Biderman's words, "Selfhood may simply be a linguistic construction, a narrative that is not defined independently by the individual, but is best explained by the storyteller." In the end, however, Biderman rejects this account of personal identity in favor of an existential account, according to which we define who we are by choosing to take action and thus define our identity. "In this sense," Biderman writes, the protagonist, Douglas Quaid, "overcomes his past identity and the idea that the self is a linguistic construction by leading the authentic life."

A problem that arises in connection with this existential account is that it seems as if there must *be* something that is *doing* the choosing, something that provides a locus for personal responsibility, much vaunted by existentialists. Unless there is some way to make sense of this, we are left with the bare conception of choices without a chooser, a notion that is difficult, if not impossible, to understand.

In "Picturing Paranoia: Interpreting *Invasion of the Body Snatchers*," I criticize political and feminist interpretations of this 1956 science fiction classic and offer a novel reinterpretation that brings out the film's concern with the philosophical significance of paranoia. I argue that *Invasion* is best understood as a film noir and that its political meanings, about which critics disagree, are not central to understanding the film or appreciating its philosophical importance. In a departure from what might be called the standard interpretation, I reinterpret features of the film that commentators usually treat as defects attributable to the studio's insistence that the story be put in a framework that gives it a happy ending. My essay attempts to convince readers that what the film seems to be about is not what it is about at all but rather reflects the filmmakers' irony. I suggest that on my interpretation, *Invasion of the Body Snatchers* is less predictable and more interesting and therefore a better film than it is according to the standard interpretation.

In "The Existential *Frankenstein*," Jennifer L. McMahon states that according to the existentialist philosopher Martin Heidegger and the psychological theorist Ernest Becker, those who fear death tend to deal with their anxiety through obsession or denial. In McMahon's words, "*Frankenstein* illustrates the anxiety that individuals have about death . . . and their desire to conquer it." In an effort to avoid the "mad scientist" clichés that cluster around the film, McMahon develops an existential picture of Victor Frankenstein, the scientist who uses technology to create and sustain life, thereby "cheating" death. McMahon finds a loss of humanity in Frankenstein's diminished capacity for considering consequences, which itself derives from his obsession to "defeat death."

The concept of humanity is multistranded, and it is open to doubt whether Victor Frankenstein's obsession in and of itself causes or constitutes a loss of humanity. One might cite infants, who have not yet developed the capacity for considering consequences, and victims of Alzheimer's disease, who have lost that capacity, as counterexamples to the thesis that such a capacity is a necessary condition of an entity's humanity. Moreover, technologically advanced robots presumably have the capacity in question, yet they do not strike us as being human simply by virtue of possessing it, so it is not clear that the capacity for considering consequences is a sufficient condition for humanity either.

Most of us are neither obsessed with death in the manner of Victor Frankenstein nor locked in a state of extreme denial but fall somewhere on the continuum between these extremes. The tenacity with which most of us cling to life and go about the business of living our lives rather than dwelling on death, even as we mourn the loss of loved ones, may reflect our belief that the continuation of conscious experience is a positive good, something to be hoped for even when some of our experiences are painful. It seems to be rational to believe that the irreversible and permanent cessation of experience is something a rational person would seek to avoid, in the absence of extraordinary circumstances. Seen in this light, there is nothing irrational or unhealthy about the wish to postpone death.

Part 2 explores extraterrestrial visitation, time travel, and artificial intelligence. In "Technology and Ethics in *The Day the Earth Stood Still*," Aeon J. Skoble uses the influential 1950s film to discuss the various roles of science and technology and their ethical implications. Commenting on the film's early sequence in which a soldier fires at Klaatu, the extraterrestrial, injuring him in the process, Skoble writes, "It's only justifiable to kill an alien who is attacking you, not one who comes in friendship bearing a gift, and . . . if Klaatu had been killed by the soldier, Gort would have killed *all* the soldiers, and maybe even destroyed Earth. Thinking about these reasons why the soldier acted badly in shooting Klaatu thus points us toward more general ethical principles about the use of force." In connection with these more general principles, Skoble offers a prudential reason for not using force ("It's not prudent to attack someone whose retaliation will be devastating or whose retaliatory capabilities are unknown"). These considerations are plausible enough, but they seem to imply that preemptive attacks are never justified, and some readers may doubt this. After all, what transpired that day in Washington DC had no precedent in human affairs. Imagine: an extraterrestrial lands his spacecraft in the nation's capital, emerges in full flight regalia accompanied by a menacing-appearing robot, and displays an instrument that looks like a weapon to a nervous, possibly inexperienced member of the military who believes, reasonably enough, that it is his duty to protect the public. Given these extenuating circumstances, it might be true that the soldier showed poor judgment or lack of self-control, but at the same time there seems to be some justification for what he did.

In his speech at the film's conclusion, Klaatu concedes that the interplanetary confederation he represents has by no means achieved perfection but only a system that works. By this he means that "at the first sign of violence," Gort and the rest of the robot police "act automatically against the aggressor." But Klaatu never acknowledges that this system might itself be tied to a fallible technology that "needs an emergency override protocol," as Kim Newman points out.[8] Moreover, viewers are asked to assume that Gort understands such notions as aggression and violence, as if their meanings were not context dependent. It would seem that any action Gort might take would require not only knowledge of the specific situation involved but also reflection and judgment and thus be far from automatic. The fact that Gort might indeed destroy Earth shows the need for "Klaatu barada nikto," the override protocol of which Newman speaks. But this also shows that Gort and his counterparts have *fallible* judgment. To take only the most obvious example, in the 1980s Ronald Reagan endorsed a military buildup, the Strategic Defense Initiative, and the abolition of nuclear weapons. How would Gort distinguish between defensive and offensive weaponry without having knowledge of President Reagan's intentions? And how could he have infallible knowledge of them, as opposed to fallible beliefs based on probabilities?

In "Some Paradoxes of Time Travel in *The Terminator* and *12 Monkeys*," William J. Devlin explains some of the paradoxes of time travel, illustrates them with a discussion of the films referred to in his essay's title, and clarifies the conceptual network that makes up our idea of time travel. Devlin identifies two types of paradox: The first is an empirical paradox, which derives from the experiences of the perspective of the time traveler. Here the interest lies in questions about what may happen to one's sense of self in time travel, the effects that changes in the past may have upon oneself, and so on. The second type is a metaphysical paradox, which derives from logical impossibilities that arise from the concept of time travel. Metaphysical paradoxes of time travel include such issues as the ability to eliminate one's own past self and the problem of finding an original cause in circular causal chains. Using *The Terminator* and *12 Monkeys* as test cases, Devlin asks whether the possibilities these films ask us to consider actually *are* possibilities, whether what we see on the screen ever *could* occur. Thus Devlin's essay raises the nagging questions, Are not the alterations that we witness onscreen, or are asked to imagine, so radical that, when the time traveler "returns" to his present, there would be no *there* there? And if this is what we are being asked to imagine, is it a logical possibility?

In "*2001*: A Philosophical Odyssey," Kevin L. Stoehr interprets *2001: A Space Odyssey* as a meditation on the nature and value of human existence. After providing readers with a tour of the film's cinematic landscape in which he makes apt comparisons with *The Searchers* (John Ford, 1956), Stoehr points out applications of the Heideggerian view that "there are no absolute or Archimedean standpoints for beings such as ourselves." He also reiterates Heidegger's rebuke to those who worship at the shrine of technology and supplements it with philosopher Hubert Dreyfus's discussion of "the existential, psychological, and moral dangers of technology in general." However, these somewhat one-sided polemics on the dehumanizing effects of technology might themselves be rebutted by anyone who has shared in the benefits of air conditioning, e-mail, or an MRI. It is as if these critics have no feeling for what the philosopher Irving Singer refers to as "the grandeur of modern technology" or "those who produce and comprehend its arcane but godlike achievements."[9] It is thus an irony that will not be lost on readers that Kubrick's dramatization of the existential hazards of an overly technological world is itself made available through the technologies of film, the VCR, and the DVD.

Scenarios in which human beings place computers, with their artificial intelligence, in control of the fates of human beings can be found not only in *2001: A Space Odyssey* but also in films such as *Dr. Strangelove, or: How I Learned to Stop Worrying and Love the Bomb* (Stanley Kubrick, 1963), *Colossus: The Forbin Project* (Joseph Sargent, 1970), and *The Terminator*. Kim Newman writes that in *Colossus* and *The Terminator*, "computer/missile link-ups . . . lead to disaster because megalomaniac machines attain a level of self-awareness that encourages them to execute self-aggrandizing schemes of world conquest."[10] The emphasis in *The Day the Earth Stood Still* is on the fallibility of human beings; the later films stress the culpability of the computer. Much of the implausibility of *Colossus: The Forbin Project* and *2001: A Space Odyssey* derives from the tendency in these films to equate rationality with computability and computability with the acquisition of factual knowledge. The assumption seems to be that a computer could surpass human knowledge-acquisition skills so completely that it could take over the world. Being strictly computational, it would be perfectly rational, which is to say unemotional (or perhaps nonemotional). Computers in these films are portrayed as making judgments about what is best, and yet, as we have seen in connection with *The Day the Earth Stood Still*, this is problematic. As long as computers are limited to gathering factual data, the norms and values that are necessary for formulating policy ends and goals will be missing. The problem is therefore to explain how normative thought, which is essentially value laden, is possible without the input of premises that are not strictly factual but evaluative. For questions of policy ends and objectives ultimately turn on what kind of society we *ought* to have and what ways of life are the *best* ways of life. These questions implicate values in a fundamental way. They are not answerable simply by acquiring more information.

In "Terminator-Fear and the Paradox of Fiction," Jason Holt uses *The Terminator* to consider not only artificial *intelligence* but also artificial *consciousness*. He calls attention to cinematic techniques that give us imaginative entry into what an artificial consciousness might be like. Holt also uses *The Terminator* as a case study to explain and solve the so-called paradox of fiction, which can be succinctly expressed by the question, How is it that people can feel fear, compassion, admiration, and other emotions for something they know does not exist? When we reflect on our emotional responses to characters in films, there is clearly something odd going on, as can be seen from the following three propositions: (1) Audiences often experience emotions (fear, pity, desire) toward things they know are not real (for example, characters in a film). (2) We can experience such emotions only if we believe that the objects of the emotions (that is, the characters) exist. (3) Audiences who know that such objects are fictional do not believe that they exist. The paradox arises because while each of these propositions seems plausible and likely to be true, they jointly imply an inconsistent set: they cannot all be true together. So Holt's question is, How is it that we feel emotions toward things that we know do not exist, when such feelings seem to depend on believing that they do exist? How can we fear the Terminator or Frankenstein's monster or the creature in *Alien* when we know that such things are not real? Unless this paradox can be solved, it would appear that our emotional responses to fiction (including science fiction and horror films) are irrational.

In part 3, "Brave Newer World: Science Fiction Futurism," four essays explore the various ways science fiction film has conceived of the future and our place in it, starting with "The Dialectic of Enlightenment in *Metropolis*" by Jerold J. Abrams, who discusses this dystopian vision of modernity gone awry. *Metropolis* is a precursor of *Alphaville, Blade Runner,* and *Dark City*—an influence on them and, viewed in retrospect, a constant reminder of them as well. Abrams provides a detailed and highly allusive account of this masterwork, often called the first feature-length science fiction film. One of the most impressive aspects of the film, as Abrams points out, is that it anticipates by almost two decades the critique of Enlightenment modernity articulated by Adorno and Horkheimer in *Dialectic of Enlightenment* (1944). For Adorno and Horkheimer, the totalitarianism, estrangement from nature, and retreat from reason that are characteristic of modernity can be traced to the Enlightenment's self-critical attack on reason itself. For the Enlightenment's conception of reason did not end with "reason's technical overcoming of nature and mythology," in Abrams's words. Instead, Enlightenment reason bit its own tail and ushered in the epistemological skepticism and nihilism that typify modernity and find expression in fascism.

As a diagnosis of the ails to which modernity is subject, however, it might be argued that the ominous strains in the Enlightenment conception of reason can be attributed to the *uses* to which technical mastery and scientific control can be put. Adorno and Horkheimer's critique conveys an attitude toward the aims, strengths, and achievements of science and technology that is strikingly similar to that in Heidegger's critique of instrumental rationality and technology. These critiques might more appropriately be directed toward the aims, limitations, and betrayals of those who control these wonderful mechanisms of reason and understanding.

Another theme discussed by Adorno and Horkheimer in their *Dialectic of Enlightenment* has important implications for popular culture. It is their view that our experience of mass media, and film particularly, is mostly passive and that this is a bad thing. Against this view, one can argue, as Richard A. Gilmore does in his discussion of the Adorno-Horkheimer critique, that this passivity may be one of the benefits of watching movies. "In going to the movies," he writes, "I both lose the sense of impinging reality and feel completely satisfied with the (frankly, rather minimal) activity in which I am engaged."[11] One can also deny the initial assumption that our experience of films is passive. Contrary to Adorno and Horkheimer's position, looking at a film, as distinct from simply gazing at the screen, is active. Normally when one watches a science fiction film, one takes notice of its style and themes. One makes connections, formulates hypotheses, seeks meanings, and asks oneself questions about what will happen next. Far from being passive, watching a film typically involves attentiveness.

When Fredric Warburg, the British publisher of George Orwell's novel *1984*, read the manuscript, he reported to his colleagues, "*1984* . . . might well be described as a horror novel, and would make a horror film which . . . might secure all countries threatened by communism for 1,000 years to come."[12] *1984* has appeared as a feature film twice, in 1956 and 1984. R. Barton Palmer focuses on the two film versions of the novel as a way of illustrating science fiction's futurism as a response to its historical context. A key insight in his "Imagining the Future, Contemplating the Past: The Screen Versions of *1984*" is that "imagined worlds hold an immense usefulness for a symptomatic analysis of the present." By providing a basis for recognizing what, at present, we both desire and fear, dystopian film and fiction offer "a startling alterity" and an impetus to further reflection.

Whereas the 1956 film conforms to the interpretation of Orwell's novel as a dystopian fantasy, the 1984 version, according to Palmer, seeks to debunk that interpretation, reflecting both the director Michael Radford's dubious ideas about Orwell's intentions and his sense that the novel's political themes were no longer relevant. Palmer connects the earlier version of *1984* to science fiction and film noir, where the dark underside of quotidian life suggests the truth of the observation of the poet Delmore Schwartz that even paranoids have real enemies. In Winston Smith's world, everyone *is* out to get him, and no one comes to his rescue. At the end of both the novel and the 1956 version of the film, Winston is conquered by the Party and betrays Julia. The last line of the novel is "He loved Big Brother." This devastating critique is diluted in the 1984 version of the film, which simply shows Winston looking after Julia and saying, "I love you." As Palmer points out, the latter screen version conforms to the heritage film, in which style predominates over content.

The link between science fiction and the noir private detective is forged in *Alphaville* and to similar effect in *Blade Runner*. *Alphaville*'s director, Jean-Luc Godard, makes no attempt to conceal the low-tech special effects and awkward (and quite possibly improvised) dialogue. Indeed, that is part of the point of this deeply self-reflexive film: to call attention to the medium itself and to let us in on some of its artifices and limitations. Alan Woolfolk's "Disenchantment and Rebellion in *Alphaville*" clarifies the existential predicament and spiritual agenda of the film, illuminating the conflict between modernity and modernism. Woolfolk invokes the sociologist Max Weber's notion that "the process of the intellectualization and rationalization of the world, of which 'scientific progress is . . . the most important fraction,' . . . does not lead to 'an increased and general knowledge of the conditions under which one lives.' Rather, it eliminates meaning from collective and individual life, leaving only the instrumental priorities of the present moment." Similar criticisms of science and technology can be found in connection with Heidegger's existentialist and Derrida's deconstructionist critiques. Against them it might be observed that what matters most is how science and technology are *used*. Obviously, the uses to which technology are put can be a boon or a bane to humankind. What is more, often benefits and costs are incommensurable: The personal computer and the cell phone have conferred undeniable benefits on their users by making for greater efficiency in everything from banking to purchasing postage stamps to filling prescriptions. But it is not obvious how to balance these benefits against the potential for invasion of privacy that centralized computer data make possible. But the problem Weber raises goes beyond these issues, for it gives rise to the challenge of living in a meaningless universe.

Rounding out part 3, Mark T. Conard discusses problems of epistemology and metaphysics and the use of Plato and Descartes in "*The Matrix*, the Cave, and the Cogito." Conard argues that *The Matrix* should not be construed as a contemporary analogue of Plato's allegory of the cave. Despite some superficial similarities, the two register significantly dissimilar epistemological and metaphysical views. Conard argues that the epistemology and metaphysics of *The Matrix* are empiricist; those of Plato's allegory of the cave are rationalist or even idealist. Conard's main point, expressed in my own terms, is to ask what the grounds are for believing that Neo's "reality" is not in fact a "meta-Matrix"—that is, just another construction or dream world manufactured by yet another level of agents. One finds this possibility even in Descartes, for on Descartes' own assumptions, it is possible that *Descartes* does not exist. What Descartes establishes, at least to his own satisfaction, is that a *thinking being* exists. But the "I" who is doing the thinking might not be Descartes. That it is Descartes, and not some other being, who is doing the thinking is just a contingent fact.

Conard does not go so far as to actually endorse Platonic and Cartesian metaphysics and epistemology but only to argue that without them, *The Matrix* lacks sufficient grounding to make its own case. He believes that something on this order is required if we are to avoid the endless repetition or duplication of the skepticism and solipsism that Plato and Descartes tried to avoid. From Conard's perspective it seems clear that skeptical questions will inevitably arise in connection with an outlook that fails to find anchoring in the a priori. Whether this assertion is itself defensible is a question much debated among philosophers.

Science Fiction Film Criticism

The science fiction film has a remarkable hold on the filmgoing public. From its origins in the late 1920s with *Metropolis* to present-day blockbusters like *The Matrix*, it has kept filmmakers and their audiences in thrall. This is not to say that everyone has succumbed to its charms. This is nowhere better illustrated than in the reaction to *Star Wars*, whose partisans think it gave much-needed impetus and a new lease on life to the genre, while its detractors describe it as, in the words of the filmmaker Paul Schrader, "the film that ate the heart and the soul of Hollywood" with its "big-budget comic book mentality."[13] Writing in his column of film criticism for the *New Republic*, Stanley Kauffmann says, "To enjoy *Blade Runner*, you need only disregard, as far as possible, the actors and dialogue."[14] The film and theater critic John Simon, always a good source of acid and amusing commentary, writes,

> These filmmakers—whether they be Kubrick with *2001*, George Lucas with *Star Wars*, or Spielberg [with *Close Encounters of the Third Kind*]—seem afraid, perhaps even incapable, of genuine feelings. Kubrick may pretend that his point is precisely the affectlessness of the space age—but, why, then, are his other films, about the past and present, similarly affectless? Lucas may evade the issue by telling a children's adventure story in which sexuality and love do not belong by definition. Spielberg elides the whole thing with a few moist glances and one perfunctory clinch. . . . [B]efore we know it, all zap and zowie breaks loose again. Anything to avoid addressing oneself to the relations of grown men and women.[15]

And to add frosting to the critical cake, consider the film writer David

Thomson on *Solaris* (Andrei Tarkovsky, 1972): "The 'enemy' on Solaris is the way the planet can generate the people that its inhabitants are thinking about. I do not mean to be snide when I say that an episode of *Star Trek* explored this theme with more wit and ingenuity, less sentimentality, and at a third of the length."[16]

Whether or not one finds these observations convincing, there is much to be learned from film critics and commentators, even those who express strong views like those I have quoted. Unfortunately, good science fiction film criticism remains in perilously short supply. Some bad science fiction film criticism is simply attributable to sheer incompetence, and some is due to the pervasive influence of various disfiguring intellectual tendencies in academic circles that are estranged from what were once the humanizing methods of the humanities. This should concern everyone who cares about science fiction film because good criticism of it—by which I mean criticism that is clear, consistent, carefully researched, cogently argued, and unclouded by dogma—is essential to the integrity of the genre. Virtually all bad writing about science fiction film reflects poor thinking about it, not only cheapening science fiction film criticism but also trivializing science fiction film itself.[17]

Since most science fiction films have compelling story lines and vivid imagery, it is easy to write about them with enthusiasm. But this does not mean that it is always done well. Here is C. J. Henderson, the author of *The Encyclopedia of Science Fiction Movies*, on *Invasion of the Body Snatchers*: "Dr. Miles Bennell ([Kevin] McCarthy) has been on a short vacation. He comes home to Santa Mira, California, rested and relaxed and ready to tackle his patients' problems anew."[18] Unfortunately, this gets things wrong from the start, for Bennell has been summoned home from a medical convention at the urging of his nurse, who tells the slightly irked physician that his office is filled with patients who insist that they must see him. The tendency to get it wrong continues to show itself as Henderson writes, "Bennell keeps assuring people that everything is okay. After all, he can't be bothered with all of this—he has a date with Becky. But the date gets put on hold when the couple's friends, the Bellichecs [*sic*] . . . discover a body in their home. . . . Only a bit of investigation is needed to find another body that looks like Mrs. Bellichec. The bodies are destroyed, and the four sit down to discuss what they have seen."[19] None of these things, except the discovery of a body in the Belicecs' home, occurs in the film. Bennell and his friends do not discover another body that looks like Mrs. Belicec, though they do discover seed pods in Miles's greenhouse that are disgorging embryo-type entities that bear a striking resemblance to Miles and Becky and to which Miles takes a pitchfork. But by the time the four become dimly aware of what is happening in Santa Mira, they don't have time to sit down and talk things over because they are too busy running for their lives. These errors would be reason enough to give this entry in Henderson's book a wide berth, but what makes it worse is the author's attitude that "I've developed an ego sufficiently large enough to allow me to present my opinions as fact and expect them to be taken at face value."[20] Henderson's presumption that his opinions should be taken at face value is inimical to the role that critical judgment should play in the evaluation of the views of all sides, including the critic's own.

Even works that make important contributions to our understanding of science fiction film can veer off in dubious directions. Kim Newman's readable but determinedly partisan *Apocalypse Movies: End of the World Cinema*, which I cited earlier, is a case in point. Newman observes that *Terminator 2: Judgment Day* (James Cameron, 1991) "is more in love with technology than the original, for after all this is a film made possible by enormous research expenditure and the resources of major corporations."[21] But this observation should raise one's suspicions, since the resources of major corporations also make possible films with messages about the risks and dangers of technology. Why would Newman single out *Terminator 2* unless he had an ax to grind? The answer is not far to seek. Compulsively dismissive of Ronald Reagan, Newman observes that "it has been argued" by critics of Reagan's Strategic Defense Initiative (SDI) that the orbital antinuclear defense program was modeled on one of Reagan's own movies, implying that Reagan lived in a fantasy world, whereas "SDI boosters like science fiction writers Larry Niven and Jerry Pournelle now like to claim" that the program had a legitimate foreign policy objective and was designed to bankrupt the Soviet Union.[22] Granted that neither contention comes from a neutral source, there is still a double standard: why is the latter treated with skepticism ("now like to claim") while the former is presented as objective fact finding and setting the record straight?

These two cases illustrate a few of the pitfalls that readers may encounter as they make their way through the large body of exposition, commentary, and criticism of science fiction film. My hope is that the essays in *The Philosophy of Science Fiction Film* will provide not only insight and argument but also tools for readers to use as they reexamine their own assumptions and raise further questions.

Notes

I am grateful to Christeen Clemens, Paul Goulart, and Aeon Skoble for their contributions to my understanding of the issues with which this introductory essay deals.

1. James O'Neill, *Sci-Fi on Tape* (New York: Billboard Books, 1997), 54.

2. John Scalzi, *The Rough Guide to Sci-Fi Movies* (London: Rough Guides, 2005), 79.

3. Kim Newman, *Apocalypse Movies: End of the World Cinema* (New York: St. Martin's Press, 1999), 142.

4. Andrew Spicer, *Film Noir* (Edinburgh: Pearson Education, 2002), 152.

5. A good discussion of noir films in which the protagonist suffers from amnesia can be found in Andrew Spicer's essay, "Problems of Memory and Identity in Neo-Noir's Existentialist Antihero," in *The Philosophy of Neo-Noir*, ed. Mark T. Conard (Lexington: University Press of Kentucky, 2007), 47–63.

6. Scott Bukatman, *Blade Runner* (London: British Film Institute, 1997), 69.

7. For an acute analysis of this problem, see Douglas C. Long, "The Philosophical Concept of the Human Body," *Philosophical Review* 73 (1964): 321–37.

8. Newman, *Apocalypse Movies*, 143.

9. Irving Singer, *Three Philosophical Filmmakers* (Cambridge, MA: MIT Press, 2004), 256–57.

10. Newman, *Apocalypse Movies*, 219.

11. Richard A. Gilmore, *Doing Philosophy at the Movies* (Albany: State University of New York Press, 2005), 135.

12. Warburg quoted in Stanley Kauffmann, *Field of View: Film Criticism and Comment* (New York: PAJ Publications, 1986), 272.

13. Schrader quoted in Peter Biskind, *Easy Riders and Raging Bulls* (New York: Simon and Schuster, 1998), 317.

14. Kauffmann, *Field of View*, 190.

15. John Simon, *Reverse Angle: A Decade of American Films* (New York: Potter, 1982), 345.

16. David Thomson, *The New Biographical Dictionary of Film*, rev. ed. (New York: Knopf, 2004), 882. Thomson does not say, but he is no doubt referring to the episode "Shore Leave."

17. Two highly amusing History News Network blog entries by Aeon J. Skoble discuss some of the relevant issues: "I Forgot My Mantra," June 16, 2005 (http://hnn.us/blogs/entries/12474.html), and "Let Him Through, He's a Scientist!" August 27, 2004 (http://hnn.us/blogs/entries/6994.html).

18. C. J. Henderson, *The Encyclopedia of Science Fiction Movies* (New York: Checkmark Books, 2001), 195.

19. Ibid., 196.

20. Ibid., x.

21. Newman, *Apocalypse Movies*, 247.

22. Ibid., 246.

Part 1

ENIGMAS OF IDENTITY AND AGENCY

WHAT IS IT TO BE HUMAN?

Blade Runner and *Dark City*

Deborah Knight and George McKnight

Blade Runner (Ridley Scott, 1982) and *Dark City* (Alex Proyas, 1998) take place in dystopic cities set in the future of what appears to be our world.[1] Both literally and metaphorically, these are dark cities. *Blade Runner* is set in Los Angeles in 2019. The city is a gloomy, rainy, commercially driven, multiethnic megalopolis composed of street-level stall vendors, abandoned downtown buildings, and huge modernist and Mayanesque complexes housing the most powerful members of society. Our protagonist, Rick Deckard (Harrison Ford), a former member of a special police squad, is coerced into taking on one more job, to kill four humanlike androids, known as replicants, who have escaped from their off-world colony and returned to Los Angeles literally to meet their maker, the inventor Tyrell (Joe Turkel). The Tyrell Corporation, which seems to control much of what goes on in the city, is engaged in the genetic engineering of replicants to work as slaves in the off-world colonies as well as in the manufacture of other artificial creatures. From Tyrell's apartment high above the city, the light from the sun is barely visible, while at street level, the sun's rays do not penetrate. Los Angeles appears to have suffered some terrific calamity that has destroyed much of the environment. Nature seems no longer to exist. There are no trees, flowers, or living animals in the city. The only animals we see in *Blade Runner* are sophisticated replications, for example Tyrell's owl. In the city, huge electronic advertisements promoting various corporate products illuminate the sky, while the congestion of the city is represented by traffic both on the ground and in the air. The off-world colonies are promoted as new lands of opportunity and adventure, suggesting that those who can afford to leave Los Angeles either have done so or are doing so.

Dark City confronts us with a sprawling metropolis perpetually trapped in the gloom of night, a nearly paranoid protagonist whose memory appears to have been erased, and an alarming, misanthropic group of extraterrestrials, known as the Strangers, who exert a mysterious control over the city and its inhabitants. It is a film that, like *Blade Runner*, combines science fiction with film noir but adds the innocent-on-the-run thriller to the mix. We follow our protagonist, John Murdoch (Rufus Sewell), as he tries to remember who he is and discover whether he is what the police suspect him of being, a vicious serial killer of prostitutes. It quickly becomes apparent that not only the police are interested in Murdoch. The extraterrestrials, as well as a dubious psychoanalyst, Dr. Schreber (Kiefer Sutherland), are interested in him as well, although their purposes are far from clear. There are several decidedly odd features of the city. One is that nightly it undergoes radical transformations: buildings spring up here and disappear there, something the city's inhabitants are unaware of because at midnight they fall into a virtually comatose sleep, while in the meantime their identities and memories might be transformed. Another is that nearly everyone John Murdoch asks about his hometown just outside the city, the seaside community called Shell Beach, claims to remember the place but has no recollection of how to get there. This leads Murdoch to realize that there doesn't seem to be any way to get out of the city, and surely it is a very odd city indeed, one with a highly developed transportation system, including a subway system suggestive of New York City's, that has no outside.[2] Murdoch also discovers odd things about himself; for instance, when time stops at midnight and the other inhabitants fall asleep, he remains awake and can observe the transformations that take place.

Blade Runner and *Dark City* are examples of a subgenre of science fiction known as cyberpunk.[3] Cyberpunk is associated with a dark vision of the near future on Earth, where humans are under the influence of electronic, informational, genetic, and other technologies, making it virtually impossible to distinguish between the real and the artificially replicated. This uncertainty applies to a variety of cases, from whether particular creatures are real—in *Blade Runner*, the question arises initially, for instance, with respect to Tyrell's owl as well as his assistant, Rachael (Sean Young)—to whether an individual's memories are veridical or implanted, as we see in connection with *Blade Runner*'s replicants as well as with John Murdoch and others in *Dark City*. In this essay, we will explore key science fiction themes and conventions as well as major philosophical issues that *Blade Runner* and *Dark City* raise, which revolve around the central question, What is it to be human?[4]

Science Fiction Meets Film Noir

Both films employ thematic conventions closely associated with film noir, for instance, a dystopic city, a mystery, a detective figure, a variety of characters with sinister motives, the uncertainty of romantic love, and a bleak, indeed fatalistic, tone. From the very beginning of *Dark City* and emerging partway through *Blade Runner,* uncertainty is a central thematic element. John Murdoch has reason to doubt himself and his memory from the moment he awakens at the beginning of the film in a hotel bathtub with a lightbulb swinging from the ceiling above him. Also in the hotel room are a bizarre medical instrument and the body of a dead woman. In the pockets of his overcoat are newspaper clippings detailing the serial killings of a number of prostitutes. Putting everything together, Murdoch imagines he must be the murderer. Given that he cannot remember anything, most particularly how he came to be in the hotel room in the first place, uncertainty characterizes Murdoch's thoughts and experiences from the first moments of the film. By contrast, Deckard initially appears to be a completely self-reliant figure, something of an outsider even though he is again working for the police to resolve problems threatening the Tyrell Corporation, and believes he is able to take care of himself while hunting down the escaped replicants. Thus, at the outset, it seems that Deckard does not suffer any sort of uncertainty about himself, his abilities, or the nature of his job. Nevertheless, toward the end of *Blade Runner,* Deckard has as much reason to doubt himself and his memory as does John Murdoch.

Early in *Dark City* and late in *Blade Runner,* our protagonists wind up on the run. John Murdoch is an innocent on the run, since he has been framed for the various murders he is suspected of having committed.[5] Innocents on the run typically become caught up in a set of circumstances they do not understand. They are not guilty of the crimes they are accused of but find themselves pursued by both the police and those who are in fact guilty of the crimes in question. Thus it is legitimate for John Murdoch to run since he cannot be sure that he is not the serial killer the police believe him to be, and moreover he needs to run in order to prove his innocence. Late in *Blade Runner,* Deckard and Rachael are also on the run, due in large measure to Deckard's change of heart about Rachael's moral status as a replicant and his own suspicion that perhaps he is a replicant as well. They are escaping an unjust regime whose notion of what counts as a genuine crime—for instance, that a replicant might wish to live as a human on Earth—is put into question by the film. In attempting to escape, both Deckard and Murdoch must guard themselves against capture while also trying to uncover the solution to the mysteries they have found themselves at the center of. What is striking about these two films as examples of noir-influenced science fiction drawing on the innocent-on-the-run thriller is that in both, the mystery in question is not primarily focused on discovering the truth behind a particular event or on answering the questions of who did what when, where, and why. Rather, for both Deckard and Murdoch, the mystery they are at the center of turns out to be a mystery concerning identity—namely, their own identities.

Blade Runner and *Dark City* exploit the idea, so prevalent in noir-influenced science fiction, that human or humanlike life can be either created at will by rogue scientists or manipulated by extraterrestrials with superior scientific powers. In both films we find the noir thematic of a figure (Tyrell in *Blade Runner*) or group (the Strangers in *Dark City*) who tries to control some of the film's central characters or even the protagonist himself. While each of the replicants is given a distinct human form, Tyrell's objective as their designer is to make their identity subservient to their primary functions. The escaped Nexus 6 models include those designed for combat, such as Roy Batty (Rutger Hauer) and Leon Kowalski (Brion James); Zhora (Joanna Cassidy), who was designed as a member of an off-world kick-murder squad; and Pris (Daryl Hannah), who was designed as a basic pleasure model. A further measure intended to control the replicants is the fact that Tyrell, in conjunction with the police, has made it illegal for replicants to return to Earth from their enslavement in the off-world colonies. Moreover, it is because of Tyrell's enormous power and close connection to the police that Bryant (M. Emmet Walsh), Deckard's police boss, can force Deckard into taking on the job of terminating the replicants. As Deckard realizes, he has no choice.

The Strangers are a collection of ashen-faced, black-robed male creatures of different ages who enjoy certain superhuman abilities, for example the ability to stop time and to rearrange physical space. They are an ominous lot, as we discover most chillingly through the figure of the youngest Stranger, a malevolent male child who repeatedly threatens John Murdoch.[6] Their primary henchman is an equally disturbing human psychoanalyst, Dr. Schreber, who we eventually discover is central to the complex experiments with implanted memories that the Strangers have been conducting on most of the inhabitants of the city. The character of Schreber is a wonderful collection of tics and oddities. As played by Sutherland, Schreber seems to always have to gasp for breath, and he is periodically found by Murdoch in public bathing houses because, it appears, the Strangers cannot abide water, so it is the one place where he is safe from them. During the course of the film, John Murdoch must learn to trust Schreber if the Strangers' plans are to be defeated.

Both Deckard and Murdoch must determine just which broader schemes they are pawns in. It is one thing for Murdoch to be pursued by the police, who suspect he is a murderer. But who are the others who are after him, and why? Deckard must reexamine his own identity when it occurs to him that if he is a replicant, he may, like Roy and the other Nexus 6 models, have a very limited lifespan. These narrative twists display the fatalistic element of film noir, where Deckard and Murdoch must struggle to regain control over their circumstances. Until they sort out the schemes they are each inadvertently part of, neither Deckard nor Murdoch fully understands what he is caught up in or the potential consequences that lie in wait. Only when they understand these things can they take action to extricate themselves. At the same time, both Deckard and Murdoch must discover who they are. For example, in the director's cut of *Blade Runner,* it is strongly hinted that Deckard is not human after all but a replicant created to kill other replicants. If Deckard believes that he is human, he is likely to make choices and take actions that he would not take if he came to believe that he was a replicant. The issue becomes focused through Deckard's increasing involvement, including his romantic involvement, with Rachael. Deckard not only knows that Rachael is a replicant, he knows what memories Tyrell typically implants in his replicants, which means that Deckard can tell Rachael about memories she has never disclosed to anyone, let alone to him. As he becomes more strongly attracted to Rachael, he realizes that he cannot complete the job Bryant assigned to him—to kill her.

Two events have particular importance as Deckard tries to discover his own identity as either human or replicant. The first event occurs as Deckard examines the various sets of photographs that the replicants use as memory devices. The replicants have implanted memories, as well as family photos provided to them to support their memories. Deckard realizes that one of his own family photos is identical to a photo possessed by the Nexus 6 replicant Leon, leading him to think that perhaps he too is nonhuman. The second event occurs as Deckard and Rachael escape from his apartment at the end of the film, when Deckard notices a tiny origami figure in the shape of a unicorn. Previously in the film, Bryant's assistant, Gaff (Edward

James Olmos), has left suggestive origami figures for Deckard to notice. The origami unicorn is particularly important to Deckard because, earlier in the film, he had a dream that featured a white unicorn. So while the unicorn is emblematically the figure protecting the virgin, and as such suggests Gaff's support of Deckard's actions to save Rachael, Gaff's origami unicorn may also strongly suggest to Deckard that Gaff knows Deckard's memories are implanted just as Deckard knows Rachael's are implanted.[7] If Deckard, too, has implanted memories, then it is a simple step to conclude that he is a replicant. The film is not decisive on this issue but strongly implies that Deckard is not human.

Eventually in *Dark City* we learn the significance of the macabre medical instrument found in the hotel room John Murdoch wakes up in. This instrument is used by Schreber to implant memories in the various people the Strangers are interested in experimenting on. Although initially it might appear that this is the weapon used to kill the dead woman Murdoch finds in his hotel room, later we learn that it was in fact intended to be used to inject Murdoch's brain with memories that would coincide with being a serial killer found in a hotel room with a murdered prostitute. What the Strangers have achieved as a result of their various scientific experiments is a close analysis of how humans act given their beliefs about their own pasts. Hence the Strangers enter the city each night at midnight, stopping the clock as they work, to restructure the memories of certain inhabitants as well as the cityscape in which the inhabitants live. The Strangers believe that memory is what is characteristically human, and thinking that, they transform human memory at will to see what the results might be. Later, Murdoch suggests that memory might not have been their best choice to discover what is quintessentially human. And he is in a good position to offer this thought, since his memory has repeatedly been completely reprogrammed by the Strangers.

Philosophical Themes

Science fiction is a genre that exploits, probably more than any other, a range of central philosophical themes and topics. Of course, various genres raise philosophical questions. The western and the crime film ask, in different ways, What is justice? The family melodrama and the romantic comedy ask, in different ways, What is love? But it is a feature of science fiction to ask such questions as, What is it to be human? What are the conditions of personal identity? What are the roles played by reason/desire/memory in human existence? Both *Blade Runner* and *Dark City* examine these key themes. Arguably, these are the sorts of questions that naturally arise from noir science fiction narratives that feature high degrees of uncertainty: uncertainty about one's identity and actions, uncertainty about who and what can be trusted, uncertainty about what is real and what is either false or fabricated.

WHAT IS IT TO BE HUMAN?

Let us start with the question of what it is to be human. *Blade Runner* introduces this theme in relation to the status of the replicants, while in *Dark City*, the Strangers perform complex experiments to try to discover what is unique about human beings because their own species is dying. They guess that whatever is unique about humans, if they could only discover it, might be used to transform themselves and ensure their continued existence.

In *Blade Runner*, both the extremely powerful, for instance Tyrell, and the comparatively powerless, for example the toy maker J. F. Sebastian (William Sanderson), have the competence to engineer lifelike creatures. The difference between the two, as creators, is that Sebastian's creations, although they can mimic human speech and movement, are really just highly sophisticated toys, whereas Tyrell's creatures are virtually indistinguishable from actual humans. Even for someone as skilled as Deckard, it is a challenge to determine, for instance, whether Tyrell's newest "experiment," Rachael, is or is not a replicant. *Blade Runner* initially approaches the question of what it is to be human from what might be seen as the opposite direction—that is, from the perspective of the replicants, some of whom, most obviously Roy, desire to live as humans. Replicants were designed to copy human beings in every way except for their emotions. From the perspectives both of the replicants' maker, Tyrell, and of the rebel replicants themselves, the objective is to be as human as possible. "More human than human!" is Tyrell's slogan describing the replicants. For Tyrell, this means producing creatures who can act autonomously and carry out the specific tasks for which they have been designed. However, fearing that replicants might develop their own emotions after a few years, the designers built in a fail-safe device—a four-year lifespan. So, for a replicant like Roy, to live as if he were human means escaping his status as a slave in the off-world colony and his preprogrammed four-year lifespan. Why should beings who are "more human than human" be treated as slaves and be forced into demeaning labor and violent actions on behalf of the actual humans who created them? While trying to contact Tyrell, Roy and Leon visit the designer who created their replicant eyes, and Roy says to him, "If only you could see what I've seen through your eyes." What Roy has seen are events too gruesome for the human imagination. The suggestion is that Roy is reflexively aware of his own response to the events he has witnessed, events he realizes go far beyond the experience of the eye designer. Here there is an implicit suggestion that Roy's response is both cognitive and emotional. The general philosophical question becomes, At what point should humanly engineered creatures such as the Nexus 6 replicants count as persons?

Deckard initially is quite clear that humans are humans and replicants are mere things. Given that at the beginning of the film he believes he is human and the escaped Nexus 6 creatures plainly are not, he has no problem taking on the task of terminating them. Because he can draw such an uncompromising line between humans and nonhumans, Deckard has no moral qualms about his role as a blade runner. For him, terminating a replicant is little different from unplugging a computer. "Replicants are like any other machine," he tells Rachael when they meet. After concluding his test to determine Rachael's status and finding that she is a very sophisticated replicant, he asks Tyrell how it could be that "it doesn't know what it is." Although replicants are so similar to humans in appearance, with their (implanted) memories, their developing emotions, and their abilities to plan and complete actions of various sorts, Deckard feels no empathy toward them, hence his initial reference to Rachael as *it*. But the character of Rachael functions as a femme fatale figure in *Blade Runner*,[8] awakening the hard-boiled blade runner's emotions and desires, and over time Deckard changes his mind about both Rachael and his role as a blade runner. Although the terminations of Leon and Zhora have little psychological impact on Deckard, the death of Pris and Roy's response to her death, not to mention Roy's own death, begin to cause Deckard to rethink his view of the replicants and his understanding of their desire to live. And when Roy, who has already killed his maker, Tyrell, chooses to save Deckard's life rather than kill him, Deckard realizes that replicants deserve more concern than he had previously imagined.

MEMORY AND IDENTITY

The experiments conducted by the Strangers in *Dark City* concern the role of memory in human identity. When the memory injections work correctly, people wake up and behave in ways that fit seamlessly with their newly implanted memories. In one striking experiment, the Strangers transform a lower-class couple into an upper-class couple. They also change the setting the couple live in by providing much grander decor and clothes. Still, when the couple awaken, it is not with the same mannerisms and voice patterns that they had before being injected. The change in memory seems fundamentally to change the individual's identity. At the opposite extreme, as in Murdoch's case, if the injection fails because it is not properly administered, there are only memory scraps left but no full-blown sense of personal identity. Murdoch remembers his wife, Emma (Jennifer Connelly), and some episodes from his childhood at Shell Beach, but little more.

How central is memory in establishing human identity? Arguably, memory is essential for coherent, ongoing action. If you can't remember who you know or what you value, how will you be able to decide what to do? Not to mention, if you don't remember who you are, how will you be able to make any practical decisions except as contingent responses to circumstances you don't fully understand? This is the situation in which Murdoch finds himself. These potential problems also suggest one reason why *Blade Runner*'s replicants have been given implanted memories. Of course, it is equally possible to imagine mere machines that are programmed to perform the sorts of tasks that Pris, Roy, and Rachael have been created to perform, something that features centrally in many science fiction films, such as the *Terminator* series.[9] But the issue in *Blade Runner* is not the issue in the *Terminator* films. It is not a question of our hero's killing a pure killing machine. Rather, it is a question of dealing with creations that mimic humans much more closely than the machines in the *Terminator* series.

What is the influence of memory implants on the replicants in *Blade Runner*? Perhaps most obviously, the replicants' memories necessarily give them a false sense of themselves. They believe they have experienced situations, for example with family, that simply never existed, given that the memories in question have been concocted by Tyrell. Their implanted memories suggest that they are much older than in fact they are—in their twenties or thirties or even forties as opposed to being technically under the age of four. Their implanted memories arguably also allow them to imaginatively project themselves into a future they will not live to see, which leads to the revolt of the Nexus 6 models. The reason to implant false memories in the replicants is arguably to make them easier to control, while their four-year lifespan is intended to prevent them from developing more complex emotional responses and personal relationships, such as we see between Roy and

Pris. Either of these would make them unpredictable and thus considerably harder to control.

Given their predicament, the replicants must rely on the memories they have, with nothing to fall back on when those memories are revealed to be false. This is the situation Rachael finds herself in when Deckard recites to her some of her most closely guarded memories. How is she to understand herself if things she thinks of as defining her turn out to be untrue—indeed, if the events she remembers turn out never to have taken place at all? The same crisis of identity would befall the inhabitants of *Dark City* if they learned that their memories had been created by the Strangers from the memories of others to persuade them that this or that happened to them in the past. In fact, just such a crisis does befall John Murdoch, who comes closest of all the characters under discussion to acknowledging that he must carry on into the future without having any clear sense as to what is true and what is false about memories that previously he has taken to be life-defining. One can clearly see the strategy of both the Tyrell Corporation and the Strangers. Both assume that a complex and coherent memory narrative will serve to establish any individual's identity, however concocted the memory narrative might be. While this might work in some cases, those who recognize that the memory narrative is false will also recognize that they cannot rely on it, and thus their identity will emerge as a pressing question.

EMOTIONS AND MEMORY

Both *Blade Runner* and *Dark City* ask us, not unreasonably, to assume that memory plays a central role in human life and individual identity. Of course, memory can be construed in different ways. It can, for example, be construed in primarily cognitive terms, for instance in terms of memories of this or that fact, place, or event. When memory is viewed in primarily cognitive terms, feelings or emotions need not be involved at all, as when we remember our zip code or telephone number or yesterday's uneventful trip to the supermarket. Memory can alternatively be construed in primarily emotional terms, such as John Murdoch's remembering how he feels about Shell Beach or that he loves his wife. Of course, the problem John Murdoch confronts is that his memories are implanted. On the cognitive side of things, nothing is veridical. There is no Shell Beach, after all. Still, *Dark City* strongly suggests that the emotional tone of one's memories is what matters. As Emma says, "I love you, John. You can't fake something like that."

Blade Runner also suggests that it is not the cognitive aspect of memory that is the uniquely defining feature of humans but rather the emotional aspect. The Tyrell Corporation seems to have anticipated that this is the real long-term problem for its replicants, hence their four-year lifespans. In the last minutes of his life, Roy displays the sort of empathy that puts Deckard, for example, to shame. Shame is hardly the response Deckard would have initially imagined he would experience with respect to a replicant. Nor did he think he would feel personal concern for a dying replicant, as he seems to feel for Roy. Although he felt nothing in the killing of Zhora and Leon, and only some concern about Pris, Deckard becomes emotionally involved in Roy's death. The way Roy's death is shot and edited means that the audience, along with Deckard, becomes sympathetically engaged with Roy's situation. Although Roy has the opportunity to let Deckard fall to his death from a rooftop, he saves him instead. Deckard scrambles away, no doubt certain that Roy still intends to kill him. But instead, Roy has realized that his own death is imminent. He confesses to Deckard, "I've seen things you people would not believe." Everything he has experienced, he realizes, is about to disappear. "All those moments will be lost in time like tears in the rain. Time to die." A dove flies from the rooftop as Roy dies. There is a shot of Deckard that is held long enough for us to consider his expression as he looks at the dead replicant. This shot then slowly dissolves into a shot of Roy. Midway through the dissolve, the images of Deckard and Roy overlap at the head and shoulder, arguably suggesting their common nature.

In both *Blade Runner* and *Dark City*, our protagonists come to new understandings of their identities, ultimately discovering that who they are differs markedly from what their initial beliefs about themselves suggested. Both Deckard and Murdoch initially believe they are human. In due course, Deckard comes to suspect that he is a replicant. Roy's death arguably brings home to Deckard that he and Roy are alike. Murdoch, upon whom the Strangers and Dr. Schreber both experiment, discovers that he has super-human powers, powers that he shares with the Strangers. Not only does he remain awake during the periods when the Strangers stop time and conduct their experiments in the city, but he also discovers, with some help from Schreber, that he has psychokinetic powers. In his final confrontation with the Strangers, Murdoch acts on the knowledge that nearly everything he previously believed to be true, both about his identity and about the world, has been concocted by the Strangers. The city, as it turns out, is not what it seems, namely a metropolis located somewhere in America. Ultimately, Murdoch discovers that the city is an artificial construct, a space station in orbit. The reason no one remembers how to get to Shell Beach is because, in fact, there is no Shell Beach; it is just an implanted memory. The reason the subway system does not go there is because, as Gertrude Stein once remarked, there is no there there. Murdoch eventually realizes that he has more to worry about than merely whether or not he is a serial killer. He realizes that he cannot rely on what he thinks he knows about his physical surroundings. Whereas in *Blade Runner,* the replicants are the created life forms acting in relation to an actual city, in *Dark City,* it is the city itself that has been created as a laboratory for the Strangers.

Both films raise important questions about personal identity that can typically be raised only in the contexts of science fiction narratives and philosophical thought experiments. Deckard and Murdoch initially believe they are human and come to suspect they are, respectively, something less and something more than human. Murdoch has a problem even more complex than Deckard's. Every time Schreber injects a new memory implant, a new person emerges from a continuing body. This is a variation on a series of famous philosophical thought experiments dating back at least to John Locke, who wondered what would happen if the brains of a prince and a cobbler were transplanted into each other's bodies.[10] Where would personhood be located in such a situation—with the body or with the brain?[11] Murdoch has a different, but related, problem, since he realizes that he exists as just the most recent in a sequence of persons inhabiting the same body, which is also the situation Emma is in, although she does not realize it.

What is undeniable is that Deckard and Murdoch are both rational agents. They are rational agents even if they acknowledge that their memories are unreliable. As rational agents, they have to choose and pursue courses of action to realize some goal or goals. They must decide between courses of action that they value and others that they do not. They both choose possible outcomes and work toward realizing them. These goals in both cases are arrived at as a result of a combination of a desire to accomplish a certain end and the capacity to project a plan into the future. Deckard's ultimate goal turns out to be protecting Rachael. Murdoch has two primary goals, which emerge after he has defeated the Strangers. One is to will Shell Beach into existence, an act that allows him to turn his psychokinetic powers to positive use and in fact brings nature back to the city by introducing both the ocean and the sun. The second, more complicated in some ways than the first, is to win back the love of Emma. Both Deckard's and Murdoch's status as rational agents is connected to an imaginative projection into the future involving their emotions, notably romantic desire. Here we see the philosophical themes of both films reconnect with the thematic and generic conventions of film noir–influenced science fiction.

Despite Murdoch's psychokinetic powers, he is not able to determine all future events. Wonderfully, he can create both Shell Beach and a means of actually getting there, but he cannot, it seems, simply will his greatest desire, to be reunited with his wife, Emma.[12] Nor, it seems, can he simply restore to Emma her memory, since by the end of the film her memory has once again been altered so that she is now Anna, someone who does not know John Murdoch at all. Here too we might think about how philosophical themes are intertwined with generic conventions in *Dark City*. Generically speaking, there is a strong narrative and emotional investment in uniting the correct romantic couple, something that is completely problematized by the facts that, by virtue of having been transformed from Emma to Anna, Anna does not know John Murdoch, and that John Murdoch is in love with a person, Emma, who used to inhabit Anna's body but who no longer exists.

Deckard is in a similarly problematic situation. He must act with incomplete knowledge although he has very strong suspicions about his status as a replicant. As a former blade runner, he is fully aware that any replicant not actually doing the will of its designers is to be terminated. So his decision to try to protect Rachael is one that he takes on knowing it brings with it a real risk to both of them. The conclusion of *Blade Runner* is dark and claustrophobic. Having noticed Gaff's origami unicorn on the floor outside his apartment, Deckard and Rachael enter an elevator in their attempt to escape the city. This is where the director's cut ends. The much more optimistic original theatrical release of the film shows Deckard and Rachael flying in a spinner (the small vehicles that travel through the air as opposed to along the roads) across fields and meadows. This, however, is not Ridley Scott's vision. The conclusion to the director's cut of *Blade Runner* is deliberately edgy and uncertain.

The two films' conclusions initially seem very dissimilar. John Murdoch, standing at the seaside with his beloved (although she does not know this) at his side and the sun streaming down on both of them, goes off into what seems a positive future, during which, if we wanted to project beyond the end of the film, he would win Anna's love. Deckard and Rachael, by contrast, are both on the run and at risk of being killed by any police officer or blade runner who detects their identities. That said, because both protagonists are rational agents, they proceed into the future with a plan of action (to win Anna, to protect Rachael). That the future is uncertain for each of them is a given. Neither can exercise anything like total control over his future. But as agents they choose to pursue their lives into an open-ended future, hoping for the best, while concerned primarily for significant others as well as for themselves.

The Question of Human Identity

The replicants in *Blade Runner* and most of the inhabitants of *Dark City* illustrate complex philosophical questions about the relationship between mind and body, as well as the role played by memory, on the one hand, and the emotions and desires, on the other, in our understanding of human life. In both films, we have central characters who have human (or android) bodies but whose memories have been created and implanted. The suggestion in both films is that memory can be unreliable, but emotions can provide good motives for action and sustain one's identity over time.

A striking theme used by both films concerns the physical evidence that apparently supports the memories of the replicants in *Blade Runner* and the various humans who have been subjected to the experiments performed in *Dark City*. *Blade Runner* restricts the sort of evidence that counts as reinforcing memories to photographs. Thus, even after Leon has killed a blade runner who is trying to conduct a Voigt-Kampff test on him at the point where he is challenged to demonstrate emotion in relation to a question about his relationship with his mother, Leon is still determined to retrieve family photographs from his apartment. As mentioned, Deckard discovers among his own collection of family photographs one that is identical to a photograph in Leon's collection. These photographs are intended to serve as physical evidence that the replicants in fact have the life experiences that their implanted memories tell them they have. Clearly, the fact that Deckard possesses a photograph identical to Leon's means that there is reason to doubt that these photographs are veridical. But both Leon and Deckard respond emotionally to their photograph collections. Leon, against his own best interests, wants to gain access to his apartment to retrieve his, and Deckard spends a bleak and possibly drunken evening looking over his. Leon never reaches the stage where he can imagine that his photographs are fakes. Deckard, by contrast, has good reason to recognize that these seeming tokens of his past have been manufactured to support false memories.

Dark City takes the theme of physical evidence for implanted or false memories quite a bit further. Not only are there photographs of events in John Murdoch's childhood: There are the slides of his childhood shown to Murdoch by his uncle Karl (John Bluthal). There is the question of Murdoch's scar, which is recorded in a photograph but not present on his body, as well as his notebook from his boyhood, which is blank when he first sees it but is filled with details when it is at police headquarters. Then there are the various large publicity signboards advertising Shell Beach, not to mention the subway map that indicates Shell Beach as the end of one of its lines, although it is impossible to find a subway train to take him there. Even the police inspector describes an accordion that was given to him by his deceased mother but cannot remember her giving it to him. *Dark City* emphasizes even more than *Blade Runner* how the slow recognition of false memories challenges the central characters to try to understand themselves and confirm just what their identities are. What exactly should we understand about agents whose bodies continue through time but whose sense of the past is at best unreliable and at worse false? The mind/body problem most often proposed in philosophy concerns a thought experiment in which a brain is transplanted between two bodies. In these films, we have something much more akin to new brains transplanted into one body.

In the end, it seems that the marks of continuity for the agents of these films are their bodies and their emotional relationships with others, and not their unreliable memories. Deckard may well not be human, and Murdoch may not be wholly human either. Despite their programming, they both have developed important intimate relationships, with Rachael and Emma/Anna, respectively. Both films suggest that memory is far less important in any decision about agency and personhood than are the emotions and the desires that prompt action. Where *Blade Runner* initially positions viewers in relation to Deckard with the implicit understanding that he is human and the replicants are fabricated, nonhuman beings, the film eventually undermines and reverses this understanding so that we come to recognize the replicants as those who embody the values we believe define what it is to be human: empathy, trust, loyalty, love. Although *Dark City* presents us with a protagonist we might believe to be a serial murderer, the film reveals Murdoch to be a quite different individual, whose future is also determined by his empathy, trust, loyalty and love. What Tyrell and the Strangers seek is not what they find. Despite their superior skills as scientists and designers, they lack the qualities finally discovered in Roy and Pris, Deckard and Rachael, Murdoch and Emma/Anna. Over and above the continuation of the body, the development of emotion leads our protagonists to care for others. The films end not with the illusion of stability, as represented by the sorts of memories implanted in the replicants and the inhabitants of *Dark City*, but with our protagonists risking the uncertainty of the future because of their emotional involvement with the women they love.

Notes

1. Although New Line Productions Inc. copyrighted *Dark City* in 1998 and first released it in United States on February 27, 1998, the prints of the film have a copyright date of 1997. *Blade Runner* was first released in the United States on June 25, 1982. The director's cut is copyrighted 1991 and was released on September 11, 1992. We must stress that we take the director's cut of *Blade Runner* to be authoritative. While we both clearly recall the original theatrical release of *Blade Runner,* complete with Rick Deckard's hard-boiled voice-over narration and an optimistic, indeed romantic, escape ending back to a seemingly untouched natural landscape, we recognize that Ridley Scott has effectively eliminated copies of the original version, which represented not his vision but rather the producer's sense of what was best to market. Here we deal with Scott's altogether darker version, stripped of the voice-over and ending not with a liberating flight from the city but rather with a much more compromised escape that leaves Deckard and Rachael very much to their own devices.

2. In fact, John Murdoch is told by Detective Eddie Walenski (Colin Friels), "You cannot get out of the city." But Walenski says he has found a way out, then jumps to his death in front of a speeding subway express.

3. For an excellent introduction to the topic of cyberpunk, see the article in *Wikipedia,* http://en.wikipedia.org/wiki/Cyberpunk.

4. Much of our recent work has been devoted to the intersection between (usually generic) films and the philosophical themes such films dramatize or illustrate. Hence our interest in narrative structure, generic influences (especially *Blade Runner*'s and *Dark City*'s indebtedness to film noir), and recurring conventions, such as the discovery of one's true identity, which we see in both of these films and in other noir-influenced science fiction films such as *The Matrix* (Andy Wachowski and Larry Wachowski, 1999), as well as the problematizing of the romantic couple relationship. On the philosophy of film noir, see Mark T. Conard, ed., *The Philosophy of Film Noir* (Lexington: University Press of Kentucky, 2006). On *The Matrix,* see our "Real Genre and Virtual Philosophy," in *The Matrix and Philosophy: Welcome to the Desert of the Real,* ed. William Irwin (Chicago: Open Court, 2002).

5. *Dark City* offers an interesting innovation in the innocent-on-the-run thriller. Conventionally, the central protagonist, who is being pursued, meets a woman he has never met before around whom the theme of trust is developed. In *Dark City,* the theme

of trust is developed around his wife Emma's affair, a betrayal of trust by Emma, which the Strangers believe will lead to Murdoch's killing her but which Murdoch realizes never took place and was an implanted memory. The theme of trust is also developed between Murdoch and Schreber and through the increasing trust Inspector Bumstead (William Hurt) experiences for Murdoch as he comes to believe his story.

6. Late in the film, we learn that the Strangers have simply taken over the bodies of dead humans. As it is explained to Murdoch, "We use your dead as a vessel." We never discover the real form of the Strangers.

7. Just what Gaff's knowledge is, and his exact relationship to Rachael, is ambiguous. After Roy's death, Gaff arrives on the rooftop and challenges Deckard, saying, "I guess you're through?"—a remark that again reinforces the symmetries between Deckard and Roy, whose dead body is nearby. Gaff also shouts to Deckard, "It's too bad she [Rachael] won't live. But then again, who does?" This episode precedes Deckard's discovery of the unicorn figure left by Gaff outside his apartment.

8. Rachael's wardrobe, with the square-shouldered suits, and her hairstyle, not to mention her laconic delivery of lines and frequent cigarette smoking, all suggest the classic femmes fatales of the 1940s.

9. While these are hardly the only examples of science fiction films featuring killing machines, it pays to contrast Deckard, for example, not to mention the replicants, with the central figures in *The Terminator* (James Cameron, 1984), *Terminator 2: Judgment Day* (James Cameron, 1991), and *Terminator 3: Rise of the Machines* (Jonathan Mostow, 2003).

10. This thought experiment occurs in chapter 27 of John Locke's *Essay Concerning Human Understanding,* various editions of which are currently available.

11. Two excellent introductions to the general problem of personal identity illustrated by this thought experiment are the following: John Perry, ed., *Personal Identity* (Berkeley: University of California Press, 1975), which includes the relevant chapter from Locke, and John Perry, *Identity, Personal Identity, and the Self* (Indianapolis, IN: Hackett, 2002).

12. That this fate will befall Emma is anticipated in the film when a Stranger calls her Anna. She says that she is not Anna, and the Stranger replies, "You will be soon—yes."

Recalling the Self

Personal Identity in *Total Recall*

Shai Biderman

Have You Ever Been to Mars?

Let's begin with what appears to be a very weird, yet simple, question: Have you ever been to Mars? I'm sorry to say that I haven't been there. Is that a valid answer? Well, yes, if you think you understood the question. But did you? Let's analyze each word to see. *Ever*, in this context, means from the time of one's birth until now. *Mars* is the known, yet hardly charted, planet at least 35 million miles from the earth. *Been to*, in this case, roughly means physically experienced, visited, or spent time at. *You*, of course, means . . . well . . . what exactly does it mean? This last word is not as easy to analyze as the others, since the question, Who are you? is actually much more complex than we may initially think. Who am I? What constitutes a self? What makes me *me*? How does my personhood, or account of a self, affect how I live my life?

These questions help to flesh out the philosophical topic of personal identity. While there are a variety of questions and issues that fall under this topic, two issues stand out as central to answering the question, Who am I? The first issue is the question of the criteria of personhood. What makes me a person? What is it about me that make me the same person over time? How do I explain the continuity of personhood? Do I say I am the same person now that I was when I was seven years old because I have the same body? The same mind? A combination of both? Something else? The second issue concerns the question of our existence as unique and individual human beings in the world. Even if I know what makes me a person, what makes me *me*, as opposed to someone else? What makes me a unique individual? Is there a characteristic, or set of characteristics, that defines me, personally?

These two issues not only are central to the topic of personal identity but stand in a direct relation to each other. That is, as the initial question (Have you ever been to Mars?) suggests, the need for criteria of personal identity is a preliminary condition for having anything sensible to say about the idea of personhood. Furthermore, the question of what makes a person a unique individual makes sense only once we determine what we mean by *person*.

Both issues play a central role in Paul Verhoeven's *Total Recall* (1990), the science fiction film adapted from Philip K. Dick's short story "We Can Remember It for You Wholesale." *Total Recall* tells us the tale of Douglas Quaid (Arnold Schwarzenegger), a twenty-first-century construction worker who has recurring dreams of being on Mars. Tired of his mundane life, he visits Rekall Inc. (a technological and commercialized laboratory that specializes in implanting artificial memories in its clients for a virtual reality experience) to take a vacation to Mars under the new identity of an interplanetary spy. But instead of being given a virtual reality experience, Quaid is introduced to a world where he is *actually* a secret agent. He discovers that his real name is Hauser and that his memories of his childhood, career, and love for and marriage to his wife, Lori (Sharon Stone), have all been secretly implanted for unknown reasons. Once he discovers this, the mundane life he once knew collapses. Attacked and chased by undercover spies on Earth (including his wife and coworkers), Quaid escapes to Mars to find his true identity, where he eventually helps the resistance, led by the mutant Kuato (Marshall Bell), to save the planet from its ruthless dictator, Cohaagen (Ronny Cox), who is threatening the lives of its inhabitants through his power and control. As we shall see in this essay, *Total Recall* tackles the topic of personal identity by raising the question, What is the identity of the protagonist? Quaid's adventure is a quest to uncover who he really is, a journey that not only raises the issues of the criteria of personhood and of individuality but also shows that these two issues are intimately related.

"If I'm Not Me, Then Who the Hell Am I?"

When Quaid meets Kuato, he tells him that he is searching for his identity: "I want to be myself again." But ever since his visit to Rekall, Quaid has had a problem: he doesn't know who he is. That is, Quaid is struggling with the two central issues of personal identity, the criteria of personhood and the question of his unique individuality. The first step, then, in achieving his goal is for Quaid to determine what it is about him that makes him the same person over time. What makes Quaid the same person he was when he was seven, and prior to his visit to Rekall, and up to the very moment that he is having a discussion with Kuato about being one's own self? Let's begin by mapping out the terrain of the criteria of personhood.[1]

When we search for the criteria of personhood, we are looking for two different things. First, we are looking for a characteristic of persons that helps us to separate persons from nonpersons. Here, the criterion serves as a necessary and sufficient condition that allows us to define something as a person where we can say that a purported person (let us use the designation A) *is* a person if and only if A displays the relevant characteristic (call that x). Second, we are looking for a characteristic of persons that helps us to determine what makes a person the *same* person over time. Here, the criterion serves as a necessary and sufficient condition of personhood that allows us to define the self over time where we can say that A at time T_1 is the same person as B at time T_2 if and only if A and B have the same characteristic, x. While both questions are central to the topic of personhood, we find that *Total Recall* concerns itself with finding the criterion that guarantees the sameness of person. In this section, we will follow the film's lead as Quaid attempts to find out who he is. The question, then, is what is this x that guarantees the sameness of person? There are various attempts to spell out what this x is. These attempts can be broken down into two different categories, the *physical criteria,* or the criteria that describe x as some sort of physical continuity over time, and the *psychological criteria,* or the criteria that describe x as a kind of psychological continuity over time.

THE BODY CRITERION

One attempt to give an account for x under the physical criteria is the *body criterion,* or the idea that a person's continuity of personhood is defined by having the same body over time so that A at time T_1 is the same person as B at time T_2 if and only if A and B have the same body. So, for instance, under this criterion, we can say that Quaid has been the same person over time from his childhood, to his time spent as a construction worker married to Lori, up to his journey to Mars to meet Kuato because he has had the same body, a physically and causally continuous body, over time. While this attempt can serve as an answer to the question of personhood, the view that a person is defined by having the same body over time is not without its problems. We need only turn to Kuato to consider a complication. When Quaid meets Kuato, he realizes that Kuato is a mutant who has no body of his own that allows us to clearly determine his personhood over time; rather, Quaid finds that there is only one body, with two identities—Kuato and George. Given this single body, we cannot physically separate Kuato from George. Thus, under the bodily criterion of personal identity, Kuato and George have the same identity. Kuato at time T_1 and George at time T_2 are the same person if and only if Kuato and George have the same body. And, since Kuato and George have the same body, Kuato and George are the same person. But this is a problem, since we find that George and Kuato are indeed two distinct persons: Kuato is the mutant leader of the Mars resistance while George is one of the soldiers who follows the leadership of Kuato. Furthermore, we find that Kuato and George have their own mental events and cognitive experiences independently of each other—Kuato remains hidden while George has his normal cognitive experiences, and George lapses into an unconscious state when Kuato reveals himself and speaks to Quaid. Thus the body criterion does indeed have its problems, and ultimately, Quaid rejects this criterion as an adequate account of personhood.

THE BRAIN CRITERION

A second attempt to give an account for x under the physical criteria is the *brain criterion,* or the idea that a person's continuity of personhood is defined by having the same brain over time so that A at time T_1 is the same person as B at time T_2 if and only if A and B have the same brain. This alternative criterion allows us to resolve the problem regarding the thought experiment of the identities of Kuato and George. Since the brain criterion is not as broad as the body criterion (as it specifies a distinct physical body part), we can isolate the brain within the body that hosts two identities to help us discern the two distinct identities of Kuato and George. That is, presumably, since each has his own head, each has his own brain. This helps to explain why only one person is conscious at a time, as the body can host only one active set of brain activities at a time, say T_1. But once that brain is no longer in an active, conscious state but is instead in a passive, unconscious state, so that the other brain becomes actively conscious, say at time T_2, then we can see that at times T_1 and T_2 we have two different persons, who are marked by different brains and their cognitive states, even though these two different persons have the same bodily host. Each person, whether it is Kuato or George, has his own personal identity that is defined by having his own particular brain (whether it is in a conscious or unconscious state at any given moment).

Like the body criterion, however, there are certain thought experiments that may serve as counterexamples to the brain criterion. Let's take one particular thought experiment as a test case, the memory swap experiment. Suppose that when the agency implants new memories into the brain and body of Hauser, they take out his old memories of being a secret agent and implant them into a new brain with a new body. The original brain and body of Hauser, with the new set of memories, becomes the one whom we know in the film as Quaid. Meanwhile, let us call the second one, who has the combination of a different brain and body, coupled with Hauser's original memories, Hauser$_1$. We thus have a dilemma. Who is Hauser, Quaid or Hauser$_1$? Most of us may be inclined to say that Hauser$_1$ is the real Hauser, even though Quaid has Hauser's brain. This inclination is due to the fact that Hauser$_1$ can recall the past experiences that Hauser has had, while Quaid cannot. While Quaid lives a life as a construction worker on Earth, unaware of anything about Hauser, Hauser$_1$ continues living a life as a secret double agent on Mars, retrieving information on both sides of the planetary war, just as Hauser did before the memory implants. The brain criterion then is not sufficient to guarantee personhood. As the thought experiment suggests, some other characteristic is needed to explain the continuity of identity. It may be the case that both the body and brain criteria fall short because they specify physical characteristics as conditions of personhood; instead, personal identity may be better defined by some psychological characteristic or set of psychological characteristics.

THE MEMORY CRITERION

One way to develop the psychological criteria approach is to address the previous thought experiment by specifying memory as the relevant psychological criterion. This brings us to the *memory criterion* of personal identity, or the idea that a personal identity is defined by having memories of past experiences, or what John Locke calls "retrograde consciousness." Under such a criterion, identity is grounded in the idea that A at time T_2 is the same person as B at time T_1 if and only if A can recall the experiences and events that B has experienced. As John Locke maintains, a person is characterized by his consciousness at the present moment and is characterized as the same person over time by his consciousness that extends backward—his retrograde consciousness—toward the past experiences through memories. The memory criterion thus answers our case of memory implants for Quaid and Hauser$_1$ by following our inclination that Hauser$_1$ is the real Hauser.

Furthermore, this view can explain our inclination to identify Hauser$_1$ as Hauser through its criterion. Hauser$_1$ at time T$_2$ is the same person as Hauser at time T$_1$ if and only if Hauser$_1$ can recall the experiences of Hauser. And since Hauser$_1$ can indeed recall the experiences of Hauser, Hauser$_1$ and Hauser are the same person.[2]

The question of memory, and whether or not having memories is a necessary and sufficient condition of personhood, is the premise of the entire film. And while Quaid does appear to search for his true identity by seeking his lost memories, throughout the film we find counterexamples that suggest that memories may not be the proper condition to guarantee personhood. The first case is Quaid himself. Who is Quaid? Under the memory criterion, Quaid's personhood is explained through his memories: since he can remember falling in love with Lori, having an eight-year relationship with her, marrying her, working in construction on Earth, making friends with his coworkers, etc., it follows that Quaid is the same person who has had those experiences. But the problem is that these experiences are all fictitious. Since all of those memories have been implanted in Hauser's brain, Quaid has no real memories. This entails that, under the memory criterion, there is no real Quaid. At the same time, there no longer seems to be a real Hauser, since the person who goes by the name of Quaid does not have any memory of the experiences of Hauser. Thus the memory criterion cannot answer the question, Who is Quaid?[3]

Not only does the memory criterion face the problem of false memories, but it also faces the problem of memory gaps, depicted in the film as both small and large. The small gap takes place within the life of Quaid as Quaid. After his psychotic episode at Rekall, an unconscious Quaid has his memory of the incident erased and is placed in a Johnny cab and sent home. When he awakes in the cab, he has no recollection of his experience at Rekall. Even when Harry (Quaid's coworker at the construction company, who is actually an agent) and his fellow agents confront Quaid about his "blabbing about his trip to Mars" at Rekall, Quaid is confused. This memory gap presents a problem for the memory criterion, since it cannot coherently account for the gap. Under the criterion, it seems as though Quaid, the person who saw Harry before and after going to Rekall, is not the same person who went to Rekall, since Quaid cannot recall the experience. Does this mean that Quaid never went to Rekall? But he did—he presented himself as Douglas Quaid, he acted like Quaid, he had the same personality as Quaid, etc. Surely it was Quaid. This makes the memory criterion appear absurd since it maintains that it couldn't have been Quaid. Likewise, as suggested in the problem of false memories, the memory criterion cannot handle the large memory gap that takes place in the film: namely, Quaid's inability to remember any of his past experiences that occurred prior to having his memory implants. As we've seen, Quaid's quest to become himself again is led by his search for his memories as Hauser. While the memory criterion maintains that Quaid is not Hauser (since he cannot recall the experiences of Hauser), notice that Quaid has certain traits that can be traced back to Hauser: Quaid has a yearning to visit Mars, he has the know-how to disarm and kill threatening opponents, he has the mental acumen to remain hidden and escape the agents chasing him on Earth so that he can safely get to Mars, and so on. But how can Quaid have these traits if he is a completely different person from Hauser? Where, in his false set of memories, does he have the experience of training as an agent or feeling the desire to go to Mars? He doesn't seem to have any such memories, and so it seems as though the only plausible explanation is that Quaid has a certain relation to Hauser that the memory criterion cannot explain.

THE PSYCHOLOGICAL CONTINUITY CRITERION

Perhaps a way to solve the tension between the identities of Quaid and Hauser is to adopt an account of personhood that is a variation of the psychological criteria account. This is the *psychological continuity criterion,* or the idea that personal identity is defined by the continuity of psychological relations over time so that A at time T_2 is the same person as B at time T_1 if and only if A inherits a set of mental features from B. Here, the set of mental features can be various types of features, such as one's personality traits, beliefs, goals, emotional dispositions, memories, etc. The faculty of memory is thus not isolated as the sole condition of personhood but is treated as a component in a larger bundle of mental features. Furthermore, it is not the case that A must inherit all of the mental features exhibited by B. Finally, it is important to see that there is not a certain feature that must be inherited throughout all temporal events for personhood; on the contrary, under such a criterion, personhood is constituted by the causal chain of inheritance of mental features as the person progresses through time. Under the psychological continuity criterion, we can overcome the problem of memory gaps, since memory is no longer the sole condition used to guarantee personhood. Whether it is the small gap within Quaid's memories or the large gap between Quaid and Hauser, we can conceive of psychological continuity as holding in each case. Quaid may not remember going to Rekall, but his knowledge of his residence, his love for his wife, his interest in visiting Mars, etc. remain and are inherited throughout the temporal events. Likewise, while Quaid may not recall being Hauser, his skills as a special agent and his fondness for Mars have been inherited, even after the memory implants.

Unfortunately, it doesn't seem as though the psychological continuity criterion can fully resolve the case of Quaid and Hauser. As Quaid searches for his past identity as Hauser, to his horror, he discovers who Hauser really is: a cold-hearted double agent who was secretly bosom buddies with Cohaagen and was the mastermind who planned to take on the new identity of Quaid so that he could infiltrate Kuato's hideout and destroy the resistance. Even if Quaid and Hauser share a certain set of mental features regarding skills as a double agent and a liking of Mars, the two seem to be so radically different that it is very difficult to say that they could be the same person. Whereas Hauser is so malicious and deceptive that Melina (Rachel Ticotin), his love interest and contact person in the resistance, considers him to be a "two-faced bastard," Quaid is a genuinely good-natured person who wants to do the right thing. Whereas Hauser wants to help build and protect Cohaagen's tyranny, even at the expense of killing innocent people on Mars, Quaid does everything he can to save the people. Whereas Hauser uses Melina in order to help Cohaagen, Quaid really falls in love with her. The differences between the two are so radical that even Quaid can't stand the thought of being Hauser. He tells Cohaagen, "The guy's a fucking asshole!" It seems, then, that even if Quaid inherited a thin set of features from Hauser, the two are so radically different in terms of personality traits, beliefs, goals, emotional dispositions, memories, etc. that they cannot be the same person.

Thus, while there have been various attempts to give a philosophical account of personhood by providing its criteria, *Total Recall* provides a series of thought experiments that raise questions and challenges to each account. We are thus left speculating whether or not we can accurately determine the criteria of personhood at all. We may be left with the skeptical position of David Hume, who, following the attempt to define personhood through specifying analytic-logical conditions, concludes that there is no enduring self throughout time. For Hume, when one attempts to find the criteria of personhood, one finds only a bundle of impressions, which do not provide the continuity required for personhood. If Hume is right, then Quaid's quest "to be myself again" is doomed to fail for two reasons: first, because there is no past self to which one can reconnect, and second, because selfhood is a fictitious entity or, at best, a linguistic convention.[4]

Existential Questions about Personal Identity

Though *Total Recall* raises the question of the criteria of personhood and seems to leave it open without espousing a definitive answer, this does not mean that we have to completely endorse Hume's conclusion that the self is a fictitious concept or a linguistic convention. On the contrary, Quaid's quest for an understanding of his self raises the second issue, the issue of the uniqueness and individuality of the self, as a central question about personhood. While the question of the criteria of personhood is typically approached from an objective, analytical, and logical perspective, the question of what makes me my own person and not someone else can be approached from the individual's own perspective in his life. By moving the focus of personal identity from the logician's laboratory of necessary and sufficient conditions to the human being's confrontation with everyday life and the world, we turn to a different landscape concerning personal identity, with a different series of questions concerning the self. These questions are existential in nature, as they focus on the individual person confronting his existence: What is it about human beings that make them unique as human beings? What is it that makes each of us unique?

This new landscape is best exemplified by Quaid's visit to Rekall. When Quaid decides to make the visit, his notion of self-identity is very secure. Aside from his bizarre dreams of being on Mars, he is very certain of who he is: he is Douglas Quaid, husband to Lori Quaid (a woman he's apparently known for eight years) and a construction worker who is looking forward to a virtual vacation to get away from the humdrum routine. Quaid seems so certain of who he is that the question of self-identity isn't even raised; rather, it is taken for granted. But then something happens. After what appears to be a botched attempt to implant an experience of a trip to Mars, Quaid radically changes his behavior. Acting aggressively and ranting about an oppressive danger, he manically screams at the Rekall employees, "You blew my cover! They'll be here any minute! They'll kill you all!" All of a sudden, Quaid appears to be a different person—his assumed identity as Douglas Quaid has exploded.

Quaid now personifies the existential attributes of a being who has been thrown into existence at birth, what the German existentialist philosopher

Martin Heidegger calls the "throwness of being." It is as if for the first time Quaid realizes that he is a being-in-the-world, a being who has been forced into the world without any prior choice and now must come to face that his self-identity is wrapped up in, and dependent upon, this foreign and unknown world. With this realization, Quaid is now alone in the world—he feels estranged and isolated since he now sees this throwness of his being, a throwness that he cannot rationally explain to those around him, since their self-identities are taken for granted, as his was prior to his visit to Rekall. His immediate response to this frightening feeling of isolation in an unknown world is thus one of confusion, angst, and despair—his behavior is erratic and chaotic, characterized by his paranoid ramblings about people who will kill them all.[5]

Though Quaid is subdued by medication, the existential challenge of dealing with the question of one's existence and one's self-identity has yet to be resolved. Through a chain of events, which include those closest to him turning on him, Quaid must now attempt to find out who he really is as a unique individual. Tormented by questions about his previous life and his previous understanding of the world around him, Quaid is driven to answer the existential questions concerning self-identity.

Following Jean-Paul Sartre, existentialism can provide an answer to the feelings of alienation, angst, and uncertainty that follow the realization of one's throwness into existence. The existentialist response focuses on authenticity, the embodiment of the individual who is completely free to create for himself his own meaningful life based on his own decisions, thereby making him a unique individual. The authentic individual is one who can at the same time acknowledge that he is a creature thrust forward into existence without a choice and realize that he is a creator who can project himself into the future with the ability to create his own self-identity. Although his past is given and cannot be changed (what existentialism calls the *facticity* of one's being), the human being can take a stance toward his past, which helps him shape his identity as he projects himself into the future. For Sartre, the philosophical view, starting from Aristotle, that the nature of the human being includes a definitive, objective essence that provides the function or purpose, and hence meaning, of one's life, is mistaken. Instead, the opposite holds true: "Existence precedes essence." The human being is thrust forward into existence at birth without a given essence, without a purpose or meaning in life. Once he comes into existence, he must decide for himself who he is—what beliefs, goals, projects, and values make him a unique individual. With this freedom to choose for himself, the human being has both the power to define his selfhood and the responsibility to be held accountable for all of his decisions. Regardless of the consequences of one's actions and decisions, the individual must accept responsibility for them, which makes one's freedom so radical that it seems more like a curse than a blessing. As Sartre puts it, "Man is condemned to be free."[6]

Many people, however, refuse to acknowledge that they are unique individuals who have radical freedom and radical responsibility. Such people fail to create their own self-identities and so fail to become authentic individuals. They are thrown into existence as beings-in-the-world and live in ignorance, or bad faith, of the fact that they are living an inauthentic life. Their bad faith is strengthened by their conformity with what Sartre calls *the crowd*, what Søren Kierkegaard calls *the public*, and what Friedrich Nietzsche calls *the herd*—the collection of human beings who form a nonindividual group mentality by following the beliefs, views, projects, etc. of one another. Each person hides from the freedom and responsibility that he has by nestling himself in the comfort and security of belonging with the others, anyone who is external to one's individuality. Those living in bad faith thus avoid the existential challenge entirely by following the steps and routines of the crowd.[7]

The existentialist would thus analyze Quaid's quest for self-identity through the ideas of authenticity, facticity, bad faith, and the crowd. Prior to his visit to Rekall, Quaid is living in bad faith. He lives the routine life that follows the steps of the crowd as he works day to day and continues to be devoted to Lori without questioning his conformity to the routine life of the crowd. Quaid's existence is treated as one of complete facticity—everything about Quaid's identity is given insofar as he does not recognize himself as a creator with the freedom to define his own beliefs, projects, or goals for himself. Even when he has a yearning to visit Mars, Lori is able to quell this desire so that Quaid eventually settles for a virtual trip instead. Quaid thus takes his existence and his identity for granted in such a way that he is living an inauthentic life. But, as pointed out above, Quaid's visit to Rekall is indicative of the human's thrownness into being. Quaid's realization that he has been thrust forward into existence without any choice is the impetus for moving outside the crowd. For the first time in his life, Quaid realizes that his self-identity is not necessarily determined by his conformity with others. Rather, he now feels isolated and estranged from the crowd—he recognizes his radical freedom and radical responsibility, and this recognition leaves him afraid, as he is alone in the world.

The existentialist response to Quaid's awakening at Rekall is to push through his paranoia and isolation (rather than falling back in step with the crowd) by overcoming such feelings of angst and becoming the authentic individual. Quaid must now see that although his past life—whether it is the past life of Hauser or the fake past life of Quaid—is a given and cannot be changed, he has the freedom and responsibility to take a stance toward the past and define himself through his future actions, beliefs, attitudes, goals, projects, etc. Now that Quaid is out of the crowd, he can take a stance toward his facticity, become a creator, and hence become a unique and authentic individual. Quaid can thus move from the objective approach of uncovering a universal essence of selfhood to the subjective, existential approach of creating an individual essence of selfhood.[8]

The Fictitious Self

While the existential view is one response to the challenge Quaid faces after his visit to Rekall, there is another view that Quaid takes into consideration. This view is the Humean skeptical view that was raised at the end of our coverage of the landscape of the criteria of personhood: the self is a fictitious entity—it is not a real thing out there in the world but only a linguistic convention. Hume's conclusion follows from the criteria of personhood. But we find that the same conclusion can be derived from our attempt to give an account of what makes a human being a unique individual. While some philosophers may follow existentialism's rejection of a universal essence, they defend this rejection by arguing that the self is fictitious. Ludwig Wittgenstein, for instance, criticizes the Cartesian view that the essence of the human being is rationality and maintains that "the I is not an object" in the sense that it is not a real entity that exists in the world. This, however, does not mean that the self is completely abolished. Rather, it is understood as a linguistic convention that is used to serve proper functions within the world of language: "The philosophical self is not the human being, not the human body, or the human soul, with which psychology deals, but rather the metaphysical subject, the limit of the world," where the world itself is limited by one's language. Similar to Hume's view, then, this approach maintains that there are no selves that really exist in the world but rather a series, or bundle, of sentences of which the concept of selfhood is a logical construction.[9]

As Jacques Derrida points out, since selfhood is not a real entity but a linguistic construction, the life of a person in the world becomes like a narrative within a text. Following his approach to analyzing philosophical texts, called *deconstruction,* a concept that is treated as an independent term in texts, such as the concept of the self, can be deconstructed to show that it is not independent but rather dependent upon other terms. For Derrida, we find that linguistically the self stands as opposing the other (as we've seen in the conceptual opposition between the two in existentialism). But such terms of the dichotomy of self and other depend upon each other and cannot exist linguistically without each other. The self thus has no objective reality—it is a story that can be understood only when we ask, Who is telling the story?[10]

We find that Quaid confronts this perspective of selfhood after his visit to Rekall. While he believes that his wife and friends are trying to kill him, Quaid must deal with the alternative view that everything going on around him is a dream. As Dr. Edgemar (Roy Brocksmith) points out to Quaid, "The chases, the trip to Mars, your suite here at the Hilton—these are all elements of your Rekall holiday." Quaid must consider whether he is living in a dream world or the people and events around him are real. But not only must he consider whether or not everything around him is real, he must also discern whether or not *he* is real. As Cohaagen points out to Quaid, "You're nothing! You're nobody! You're a stupid dream! Well, all dreams come to an end." Quaid's selfhood, as something real, is thus in jeopardy. Following the encounter with Edgemar, Quaid's self-identity as an agent on Mars, as well as the notion that the self is a real entity, is in danger of collapsing: "What's bullshit, Mr. Quaid? That you're having a paranoid episode triggered by acute neurochemical trauma? Or that you're really an invincible secret agent from Mars who's the victim of an interplanetary conspiracy to make him think he's a lowly construction worker?"

With selfhood in jeopardy of being a fictitious entity, Quaid's selfhood may simply be a linguistic construction, a narrative that is not defined independently by the individual but is best explained by the storyteller. We find that the other comes into play in providing an account of Quaid's identity, as those around him in this dream world have their own conceptions, their own stories, of who Quaid is. For instance, Edgemar sees Quaid as a patient at Rekall undergoing a paranoid episode; Cohaagen sees Quaid as a fictitious persona that is preventing the return of Cohaagen's buddy, Hauser; Melina sees Quaid initially as a liar who "used her to get inside" the resistance but later as the unique individual she still loves. Under this notion of the self as a linguistic convention, the entire film becomes a series of different narratives that construct the idea of Quaid's selfhood. With the distinction between reality and dream blurred, it points to the idea that there is no real self no matter in what context we speak of the self, whether it is the self portrayed in a narrative, film, or real life.

Ultimately, however, we find that Quaid's actions defy this approach to self-identity. When Quaid kills Edgemar and spits out the pill that is said to take him out of this dream world, Quaid makes the first decision that leads him on the path to following the existential response of creating his own authentic individuality. As Kuato points out to Quaid, "You are what you do." Quaid follows Kuato's advice that self-identity is not completely dependent upon one's past, one's memory, or the collection of views others may have of one but is also dependent upon one's own actions, beliefs, goals, projects, etc. Even when he is confronted with the fact that his past self can be understood as a causal continuity that is tied to Hauser, the deceptive and malicious bosom buddy of Cohaagen who deceives those around him and kills innocent people, Quaid follows the existential path. He takes an attitude toward his facticity by rejecting his past actions as completely defining him and chooses to create a new identity through his future actions. He becomes a savior to the people of Mars as he overthrows Cohaagen's tyranny and starts the reactor that allows air to flow through the civilization. In this sense, Quaid overcomes both his past identity and the idea that the self is a linguistic construction by leading the authentic life.

Notes

1. Before mapping out the attempts to provide the criteria of personhood, we may note that traditionally, the idea of a coherent, steady, objective identity has been a prerequisite to any philosophical discussion of human nature and the nature of the world. Plato, for instance, provides a theory of the (eternal) soul that becomes a crucial component to his philosophy. Since we are essentially souls, which are a set of Platonic Forms (universal, abstract, and enduring entities), the aim of life is to achieve philosophical knowledge of the Forms so that we can continue our real existence as Forms. Likewise, René Descartes suggests that the essence of the self as a rational thinking thing is the only thing of which we can be certain and so lays the foundation for our epistemology and further building blocks of our system of knowledge. Immanuel Kant expands on this idea of the self as a thinking thing by considering human beings as essentially free and rational agents who should follow their rational duties to shape their will into the good will and follow the universal moral laws. Traditionally, the self-present, freely acting

subject has been considered the pivot linking historicist and transcendental accounts of Western political and psychological experience.

2. See John Locke, *An Essay Concerning Human Understanding*, ed. Peter H. Nidditch (Oxford: Clarendon Press, 1975).

3. It is important to note that the discussion concerning the view that memory is the criterion of personhood is much broader and deeper than the scope of this paper. For instance, several authors raise challenging questions and problems with the criterion through the use of thought experiments. To read further on this discussion, see Thomas Reid, "Of Mr. Locke's Account of Personal Identity," in *Essays on the Intellectual Powers of Man* (Cambridge, MA: MIT Press, 1969); Thomas Nagel, "Brain Bisection and the Unity of Consciousness," *Synthèse* 22 (1975): 396–413; Sydney Shoemaker, "Personal Identity: A Materialist's Account," in *Personal Identity*, by Sydney Shoemaker and Richard Swinburne (Oxford: Blackwell, 1984); and Sydney Shoemaker, "Are Selves Substances?" in *Self-Knowledge and Self-Identity* (Ithaca, NY: Cornell University Press, 1963).

4. See David Hume, *Treatise of Human Nature*, ed. L. A. Selby-Bigge (Oxford: Clarendon Press, 1975).

5. See Martin Heidegger, *Being and Time*, trans. Joan Stambaugh (Albany: State University of New York Press, 1996).

6. See Jean-Paul Sartre, *Existentialism and Human Emotions*, trans. Bernard Frechtman (New York: Philosophical Library, 1957).

7. See Friedrich Nietzsche, *Beyond Good and Evil*, trans. Walter Kaufmann, in *Basic Writings of Nietzsche* (New York: Modern Library, 1968); Friedrich Nietzsche, *On the Genealogy of Morality*, trans. Maudemarie Clark and Alan J. Swensen (Indianapolis, IN: Hackett, 1997); Søren Kierkegaard, *Fear and Trembling and Repetition*, trans. Howard V. Hong and Edna H. Hong (Princeton, NJ: Princeton University Press, 1983); and Søren Kierkegaard, *The Concept of Anxiety*, trans. Reidar Thomte and Albert B. Anderson (Princeton, NJ: Princeton University Press), 1980.

8. The existential idea according to which one has to take responsibility and create his own meaningful, authentic life has many implications and can be taken in many different directions. One such direction is that of moral psychology, which asserts that being-in-the-world in an authentic way is nurtured by the empirical understanding that human beings define themselves from a given matrix of psychological patterns, and it is this given psychology that evolves into the ability to give meanings to behaviors and ideals. The moral philosopher Bernard Williams claims, for instance, that morality and moral integrity are rooted in this realization. See Bernard Williams, *Problems of the Self* (Cambridge: Cambridge University Press, 1973), esp. "Personal Identity and Individuation."

9. See Ludwig Wittgenstein, *Tractatus Logico-Philosophicus*, trans. D. F. Pears and B. F. McGuinness (London: Routledge, 1961), esp. 6:641; Ludwig Wittgenstein, *Philosophical Investigations*, trans. G. E. M. Anscombe, ed. G. E. M. Anscombe and R. Rhees (Oxford: Blackwell, 1953); and Ludwig Wittgenstein, *Preliminary Studies for the*

"Philosophical Investigations," Generally Known as the Blue and Brown Books (Oxford: Blackwell, 1958).

10. See Drucilla Cornell, *The Philosophy of the Limit* (New York: Routledge, 1992), and Irene E. Harvey, *Derrida and the Economy of Différance* (Bloomington: Indiana University Press, 1986).

PICTURING PARANOIA

Interpreting *Invasion of the Body Snatchers*

Steven M. Sanders

> That way madness lies.
> —William Shakespeare, *King Lear*

To all appearances, *Invasion of the Body Snatchers* (Don Siegel, 1956) is a paean to individuality and a warning of its imminent loss. Human-size pods appear in the California town of Santa Mira and begin duplicating the bodies of the residents, absorbing their minds while they sleep. Rushing into action with his growing realization that Santa Mira is being taken over by the pods, Dr. Miles Bennell (Kevin McCarthy) shows a resilient defiance in his perseverance against all odds. Evidently, we are meant to understand what it means to believe in something and fight for a cause.

Most critics have interpreted *Invasion of the Body Snatchers* as a reflection of the political and social anxieties of 1950s America. But the film's philosophical message transcends its mid-twentieth-century politics and sociology. We can see this if we approach *Invasion of the Body Snatchers* as a film that erects a drama of noir paranoia on its science fiction scaffolding. Its flashback structure with voice-over narration, unusually angled shots, scenes of claustrophobic darkness, crisply rendered dialogue, and sense of sinister purpose and impending doom are characteristics of films of the classic film noir cycle (1941–58).[1] The mise-en-scène evokes small-town insularity and touches the quick of the noir sensibility with its emblematic sequence in Miles's greenhouse where Miles and Becky Driscoll (Dana Wynter) and their neighbors Jack and Teddy Belicec (King Donovan and Carolyn Jones) come face to face with the pods and their worst nightmare. The incremental increase in Miles's problems and confusion typifies the predicament of the paranoid noir protagonist and reflects the deceptive appearances and unstable reality of the noir universe itself. In order to develop this line of thought, I need to explain the basis of my dissatisfaction with two alternative interpretations of the film.[2] I shall argue that they are unable to explain the film's philosophical significance and continuing relevance, something that my own interpretation is designed to provide.

People Are Pods

In an interview with Guy Braucourt published in 1972, Siegel said, "But let me repeat that all of us who worked on the film believed in what I said—that the majority of people in the world unfortunately are pods, existing without any intellectual aspirations and incapable of love."[3] And in a 1976 interview with Stuart M. Kaminsky, Siegel reiterated, "People are pods. Many of my associates are certainly pods. They have no feelings. They exist, breathe, sleep."[4] Even allowing for a certain amount of exaggeration in these remarks, Siegel seems to be entirely unsympathetic with the idea of the basic worth of common humanity. This attitude must have worked something of a hardship when it came to interpreting Jack Finney's *Collier's* magazine serial *The Body Snatchers,* with its loving evocation of the small town where he grew up and the charms of the commonplace.[5] Another ingredient in the unstable mix was the scriptwriter Daniel Mainwaring, whose left-wing despair over America in the 1950s propelled the original script's unhappy ending, one that is not found in Finney's serial or in the later book version.[6] Bringing things to a simmer was the film's producer, Walter Wanger. Negative audience reaction at prerelease screenings of *Invasion* convinced him that significant postproduction editing was needed to prevent the film from being dismissed as a B-grade science fiction thriller with limited box office appeal. This led the studio to release a version with a prologue and epilogue in the hope of achieving some subtlety and nuance. These additions were written by Mainwaring and directed by Siegel, though their involvement did not prevent them from deploring these alterations to the original narrative structure. What is remarkable is not that the chemistry among Siegel, Mainwaring, and Wanger worked so well but that it worked at all. Accordingly, I shall refer to the filmmakers, in the plural, to indicate the collaborative nature of the film.

In the studio version, the story is narrated by Miles in flashback, framed by a prologue and epilogue in the hospital emergency room where Miles is trying to get someone to listen to his warning that aliens are taking over Santa

Mira. Siegel has always insisted that he wanted the story to run in sequence, without the framing narrative and flashback. Siegel hated the way the studio put the story into a flashback frame, with its upbeat ending. In his interview with Kaminsky, he said he was "very much against" the frame because "it lets you know right away that something unusual is going to happen. If you start, as I wanted to, with McCarthy arriving in the town of Santa Mira, it reveals itself slowly, we understand why McCarthy can't readily accept the terrible thing that appears to be happening. And the dramatic impact of the film is reduced with the epilogue. I wanted it to end with McCarthy on the highway turning to the camera and saying, 'You're next!' Then, boom, the lights go up."[7] Instead, the scene ends with the camera tracking up and back as the shot of Miles on the highway transitions into the frame story with Miles in the emergency room.

From the outset it has been de rigueur to deplore the studio-mandated inclusion of the frame story and voice-over narration. Critics and commentators complain that the opening and closing sequences demanded by the studio, Allied Artists, "have the effect of reducing the film's immediacy and diluting the subject matter," in the words of the journalist Woody Haut.[8] But I shall argue that it is precisely the prologue, which dissolves into Miles's voice-over narration, so indicative of the deterministic fatality of film noir, and the epilogue, which renders the film's ostensibly happy ending curiously moot, that give *Invasion* its cachet. As Miles tells his story, we watch and listen, like Dr. Hill (Whit Bissell), the psychiatrist who's sizing him up for a straitjacket. Without the frame, we lose the close-up of Miles at the film's end, in all his paranoid fear, nearly collapsing in relief as Dr. Hill, who has begun to take him seriously, orders the police to block all highways in and out of Santa Mira and summons the FBI. The sudden shift in Dr. Hill's generally skeptical attitude toward Miles's story itself adds a discordant note to the narrative. Whatever Siegel and Mainwaring may have thought, the film deftly deploys its prologue and epilogue to sustain interest and suspense. These additions do not merely preserve but actually heighten the film's shock quotient by providing a far more ambiguous, complex, and interesting ending than the film would have had without them.

A Quotidian Verisimilitude

Mainwaring and Siegel use setting, circumstance, and humor to break up and reduce the gravity of *Invasion*'s dramatic events. One of the ways they do this is by means of a realistic portrayal of small-town life, depicting Santa Mira in all its "quotidian verisimilitude."[9] Grimaldi's roadside vegetable stand, the brightly lit outdoor hotdog stand, Lomax's gas station, and the triangular park in the center of downtown Santa Mira are related to the story not only literally but also symbolically. They come to express the meaning of the film itself: that everyday life, even the most ordinary, is essentially unstable and potentially verging into darkness and disorder. Thus *Invasion*'s depiction of the ordinary does more than provide the film with a realistic grounding with its pitch-perfect evocation of mood, its accumulation of period details, and its inimitable style. The film exploits the contrast between the events of Miles's everyday experience and those that his voice-over narration, with its growing paranoia, takes as its point of departure.

The outdoor barbeque and greenhouse sequences where Miles and Becky are entertaining their friends Jack and Teddy deploy the entire panoply of the film's elements and moods. Miles returns home to find Jack in the backyard at the grill struggling to get the charcoal to catch fire. Jack says to Teddy, "I need a martini. . . . I'm going to pour it on the charcoal. I can't get this stuff to burn."[10] Miles takes the martini from Teddy and hands it to Jack with the friendly admonition "For drinking purposes. . . . I'll get you something to start it." Miles enters the greenhouse, where he takes a can of starter fluid from a shelf. After Jack uses it to get the fire going, Miles returns to the greenhouse to put the can back on the shelf. As he is about to leave, he hears a pop and turns to see what it is. The development of tension has begun, for we have already been shown shots of giant seed pods foaming and opening to chords of ominous music. A form emerges from one of the pods to crackling sounds and more ominous chords. A terrified Miles, witnessing the vaguely human body disgorging from the shell, shouts, "Jack! Jack!" It's during this scene that Miles speculates that "somebody or something wants this duplication to take place."

Through the powerful image of the foursome starring in horror as the pods form themselves into duplicates before their terrified eyes, we come to know what fear and terror they must be feeling. We want to turn away, to avert our eyes, but we cannot, so we watch as Miles picks up a pitchfork. As he plunges the instrument into the chest of the Miles duplicate, we cut quickly to the telephone in the hallway, which has started to ring. The continuity script reads, "The pitchfork enters the chest with a rubbery-sounding thud. The phone suddenly rings."[11] It is this fast cut that gathers up all the intensity of the previous scene and now disperses it into the scene that follows, which begins as Miles grabs Becky and tells her, "We're getting out of here right now."

Pods and Politics

The critical literature on *Invasion of the Body Snatchers* has been dominated by the idea that the film is a Cold War allegory. Such a reading attempts to reconstruct the 1950s political and cultural climate in which the film was made. Peter Biskind's influential interpretation in his book *Seeing Is Believing* is a representative example. Biskind classifies *Invasion of the Body Snatchers* as a radical right-wing film. He writes, "Indeed, the red nightmare was so handy that had it not existed, American politicians would have had to invent it. Movies did invent it, and it served somewhat the same purpose in Hollywood as it did in Washington. More often than not, the Communist connection was a red herring."[12] Curiously, Biskind fails to mention the concerns that fostered the red scare in the first place: the Soviet Union's postwar expansion in Eastern Europe, its testing of an atom bomb, and the invasion of South Korea by Communist North Korea. He writes as though he believes that by simply omitting the names of the Soviet spies Klaus Fuchs and Julius and Ethel Rosenberg, the perjurer Alger Hiss, and other pro-Soviet fellow travelers, he will eliminate any grounds for suspicion one might have about communism's designs on the United States. The extent of the involvement of the Communist Party leadership in Moscow with its followers in the United States has been much debated, in part because of the availability in recent years of the Venona cable transmissions between the two. This is not to deny that the film has a political dimension. As Michael Paul Rogin writes, "Don Siegel made [*Invasion of the Body Snatchers*] in protest against pressures for political and social conformity."[13] But even if this is true, nobody has established whether *Invasion* is a protest against the political and social conformity called for by right-wing anti-Communists or that demanded by pro-Soviet collectivists, as Rogin himself concedes. Whether we say that the pods represent communism or McCarthyism or, indeed, the power structure that dominated Hollywood itself, their threat to autonomy and personal identity takes us well beyond the political conflicts of the day.

It is worth pointing out that Biskind's fanciful misreading of *Invasion* in his essay "Pods, Blobs, and Ideology in American Films of the Fifties" is accompanied by fabricated dialogue. He wants to show that "in right-wing films, both cops and doctors, the twin pillars of the centrist authority, are vilified."[14] In this connection, he cites the following exchange between Jack and Miles after Miles has been summoned to Jack's house to observe the duplicate of Jack on his pool table: "His [Bennell's] pal [Jack Belicec] asks him, 'Would you be able to forget you're a doctor for a while?' Bennell: 'Yes.'"[15] The problem with Biskind's reconstruction is that Miles does not say "Yes" in the scene in question. Instead, Miles asks, "Why?" And when Jack says, "I don't want you to call the police right away," Miles's response is to say, "Quit acting like a writer; what's going on?"[16] When Biskind discusses this scene in *Seeing Is Believing*, he writes that "Miles agrees."[17] Since this version predates the essay I have been citing, one must wonder what made Biskind up the stakes. He still wants to show that "in this film, the docs are sick and the cops are criminals."[18] But this fails to account for the fact that the only person who ultimately believes Miles is a "doc," the psychiatrist Dr. Hill. I will come back to this point when I discuss the filmmakers' use of irony in the epilogue.

The Black Misogyny of Siegel's Movie?

The inaccuracies in detail that I have mentioned above are not the only ones that have led to flawed interpretations of *Invasion of the Body Snatchers*. I now want to turn to another that has been made in connection with a feminist interpretation of the film. In order to do this, I will ask readers to recall the famous scene in which Miles and Becky seek refuge in a tunnel as they hide from a mob of pod people who are pursuing them with terrifying single-mindedness. They conceal themselves as the mob rushes into the tunnel and, failing to find the pair, moves outside to search the surrounding hills. Exhausted, Miles and Becky splash water on their faces from a small pool inside the tunnel. Suddenly, they hear a lovely Brahms lullaby in the distance. Miles leaves Becky momentarily to seek out its source, hoping it is being played by real people and not pods, who would have no need for music. Since the mob has already searched the tunnel and not found them, Miles tells Becky to wait there, where she will be safe: "Stay here and pray they're as human as they sound." The lullaby turns out to be a radio broadcast that the pods are listening to while they wait for a weather report as they load more seed pods onto a truck. When Miles returns to the tunnel, takes Becky in his arms, and kisses her, he realizes she has fallen asleep and has become a pod. It is a scene of great import, for in that moment Becky has been drained of the specialness that made her a person, and Miles's hopes for a life together collapse in that moment of recognition. A close-up shot of Becky with her expressionless face and vacant eyes is followed by a close-up of Miles. The reverse shot is actually part of the perspective not of Becky but of how Miles thinks he must seem to Becky and ultimately how he sees himself. The scene is matched in its intensity by the scene that follows, which shows Miles fleeing from the tunnel to avoid capture by the mob, now in full pursuit as Becky shouts, "He's in here! Get him!" In voice-over, Miles says, "I'd been afraid a lot of times in my life, but I didn't know the real meaning of fear until I kissed Becky."

In "Women and the Inner Game of Don Siegel's *Invasion of the Body Snatchers*," Nancy Steffen-Fluhr reports that when her class heard this line during a screening of the film, they burst into laughter.[19] She comments, "I was taken aback; then I was taken over. For in that laughter, implicit and complete, was a critique all the more perceptive for its innocence. They had seen the film more honestly than I had. What, their laughter asked, what *is* so scary about Becky? What *is* so scary about a kiss?"[20] It is tempting to reply that what is so scary about Becky is that she has become a pod, intent on Miles's destruction. In fact, this is how anyone who has watched the film without ideological blinders would reply. It seems clear that Steffen-Fluhr has approached the film with an agenda, for she concludes, "Although the plot asserts otherwise, when he leaves Becky (and thus leaves a precious part of his own identity), it is *he* not she, who has chosen to become an alien, a depressive, passionless drone," and thus, she adds, "the black misogyny of Siegel's ending emerges all the more clearly."[21]

As disquieting as these remarks are, those Steffen-Fluhr makes next are even more alarming: "Miles spends almost as much time exhorting Becky 'not to get involved with a doctor' as he does ogling her cleavage." This is alarming for two reasons. First, because it is intended to expose *Invasion*'s secret meaning: "It is his burgeoning intimacy with Becky, not the burgeoning pods, which is the hidden source of his fear."[22] If you watch the film reasonably attentively, or read the continuity script, you will see that Miles makes only two references to Becky's involvement with him (as a doctor), not "over and over," as Steffen-Fluhr insists. More important, Steffen-Fluhr fails to disclose the context in which Miles makes these remarks. The first occurs at shot 33 of the continuity script. The remark is immediately preceded by a reminiscence in which Miles reminds Becky that he wanted them to get married. When Becky says, "Just be thankful I didn't take you seriously," Miles replies, "*You* be thankful. I found out that a doctor's wife needs the understanding of an Einstein and the patience of a saint."[23] Seen in this context, it is obvious that Miles is reflecting sadly on the demands of his profession and the strain it placed on his ex-wife. Miles's second comment about Becky's involvement with a doctor occurs at shot 164, when he says, "Well, take my advice and don't get mixed up with a doctor . . . they're seldom at home." His remark is immediately followed by Becky's question: "What would you say if I told you I already was mixed up with a doctor?" Miles replies, "I'd say it was too good to be true."[24] I leave it to the reader to decide whether these remarks are indicative of "a man's terror of falling helplessly in love," as Steffen-Fluhr would have it.[25]

The second reason why Steffen-Fluhr's comment above is alarming is that she thinks there is something questionable or objectionable about a man who looks or glances longingly at the woman he has fallen in love with. If there is something objectionable about *this,* we are all in danger of becoming the pods that Siegel's film warns us against. Unwittingly, Steffen-Fluhr has fallen into the very trap she thinks she has set.

That Way Lies Madness

Steffen-Fluhr's belief that the laughter of her class was indicative of their having seen the film more honestly than she had makes an odd kind of sense because she believes that feelings and hunches are as reliable routes to the truth as reasoning and deliberation. She writes, "In the end, however, [Miles] suffers for his treason. Helpless, in a puddle of tears, out of control, vulnerable, he is utterly dependent for his salvation upon the male Authorities, personified by J. Edgar Hoover and his F.B.I. The psychiatrists who attend him are indistinguishable from Doc Kaufman [Santa Mira's psychiatrist]. Both are committed to a disbelief in the reliability of hunches, of feelings. They will accept only palpable, logical proof."[26] Steffen-Fluhr leaps to the conclusion that both doctors are committed to a disbelief in the reliability of hunches and feelings *in general* because *in this case* they have not been given evidence to support Miles's incredible tale and are therefore unwilling to take Miles's claims at face value. Her inference is thus a faulty generalization. In addition, it is narratively off the mark. Until the very end of the film, Drs. Bassett and Hill have no evidence whatsoever that Miles's account is true or even well supported. Why should feelings and hunches dictate their actions here, overriding logic, evidence, and deliberation? Does Steffen-Fluhr believe that they should have called out the National Guard on a *hunch* that Miles might be right?

The antirational element in Steffen-Fluhr's position is also on display when she holds it against Miles that he is willing to cut short his drinks and dinner with Becky after he receives a call from his answering service with Jack's urgent request to get over to his house right away. She writes that Jack's call "serves to emphasize Miles's essential puritanism. He immediately stops the music, snuffs out the candle, and prepares to leave. Work before pleasure. . . . When duty calls, emotions will have to get out of the way."[27] As a description of what actually occurs in the scene, this goes wrong in just about every way a description can go wrong. The music comes from the restaurant's jukebox, which Miles does not shut off. There is no candle to snuff out since Miles takes the phone call at the bar, which has no candle. And his emotions are powerfully engaged by this message of distress from Jack. One cannot help but notice the false alternative embodied in Steffen-Fluhr's account, as if duty must come into conflict with the emotions and cannot instead be reinforced by them. As a doctor, Miles thinks his professional calling takes priority over having martinis and dinner with Becky. This reflects his professional commitment to provide medical care, but it does not conflict with his emotional attachment to Jack. You may fault Miles for failing to call Jack before leaving the restaurant, but what would Steffen-Fluhr have the physician do? Tell his good friend and would-be patient, "Not now, Jack, I'm dancing cheek-to-cheek with my former college sweetheart"?

Incidentally, this scene gives the lie to those who say, with Biskind, that the film is a right-wing attack on the police, doctors, and other authority figures. What could be a stronger endorsement of the doctor's commitment and professional integrity than his willingness to disrupt a much anticipated night out with Becky to come to his friend's aid? Furthermore, earlier in the film, when Miles is asked by Becky's cousin, Wilma Lentz (Virginia Christine), if she is going crazy because she insists that her uncle Ira simply isn't her uncle Ira, though there is no discernible difference in his appearance, behavior, or memories, Miles refers her to a specialist, Dan Kaufman (Larry Gates). There is no right-wing animus against medical authorities here.

Invasion's signature sentence is "You're next!" But surely "That way lies madness" runs it a close second. It is Becky, not Miles, who utters the line when Miles suggests that he tuck her in, after he has brought her home at the end of their abortive date, thereby intimating a sexual advance. This gives the lie to Steffen-Fluhr's assertion that it is Miles, of all people, who is afraid of sex.

These weaknesses are symptomatic of the failure of both this feminist approach and of Biskind's political interpretation to engage substantively with the philosophical issues raised by *Invasion of the Body Snatchers*. The kind of interpretive closure they impose on the film is inimical to seeing what is best in it. I will sketch an account that allows us to see *Invasion* as reflecting not merely the time and place of its origin but the attempt of its makers to provide a framework by which we might think about conformity, struggle, individuality, death, and the meaning of life. If we look beyond the specific political and sociological context in which the film was made, we will be able to identify a philosophically salient element that goes beyond *Invasion*'s science fiction framework and 1950s atmosphere. We will be able to explain why it retains its interest more than a half-century after its release, a fact that the more narrowly focused interpretations have difficulty explaining.

Inflections of Incoherence

Invasion of the Body Snatchers satisfies on both the commercial and critical levels. As an example of science fiction suspense, it hews closely to narrative conventions and relates its story line with straightforward efficiency, assuring audience identification. The film is as suspenseful as 1950s science fiction can get, and its action terminates in an ending designed to provide emotional closure. But *Invasion* is also a clever piece of film noir, with its bewildered protagonist in jeopardy and its femme fatale. The studio-imposed framing device that provides the film's voice-over narration and flashback structure is an indication of the film's noir conventions. To viewers for whom *Invasion* was always more than a simple science fiction story, the film functions as a meditation on its central preoccupations, the threat to personal identity and the loss of what makes persons special. But it does this by indirection, for, as I will now show, Miles's account of pod metaphysics is inflected with incoherence. Although the magazine serial on which the film (and subsequently the novel) is based took some pains to try to explain the most pressing difficulties, these are largely ignored by the screenwriter and director, and this leads me to believe the omissions were intentional. But intentional or not, by summarizing what is so philosophically unsatisfying about the nature and mechanisms of pod duplication, I will be in a position to explain why this very incoherence is a virtue of the film and figures centrally into the significance of the narrative. Let us look at four problems that bring out these difficulties.

(1) The *problem of multiple duplicates* is made explicit in the novel when

Miles and Jack discover some pods in Miles's basement and watch them develop. Jack asks, "What happens to the original when the blanks duplicate a man? Are there two of them walking around?" Miles says, "Obviously not, or we'd have seen them. *I* don't know what happens, Jack."[28] The problem here is that if the pods are duplicating people throughout Santa Mira, then something must happen to the bodies of those they duplicate; otherwise Santa Mira would be filled with doubles of everyone overtaken by the pods or, what is worse, with corpses.

This problem is elided in the film. When Miles, Becky, Jack, and Teddy discover the pods forming duplicates of them in Miles's greenhouse, Becky asks, "But when they're finished, what happens to our bodies?" Miles replies, "I don't know. When the process is completed, probably the original is destroyed or disintegrates." The destruction or disintegration of bodies is something observable, but the film does not show this happening. Instead of answers we are left with mysteries.

(2) The *problem of the location of the pod* that duplicates a person during what Finney's novel calls the "changeover" process is well illustrated in two key sequences. When Becky becomes a pod in the tunnel, where does the pod that duplicates Becky's body come from? Since a pod presumably must be in close proximity to whatever it duplicates—the book says that the pods "imitate and duplicate whatever life substance they encounter"[29]—and since it is clear that no pod was brought into the tunnel while Miles and Becky were there together, how could the duplication have taken place at all, much less so quickly?

(3) In addition to this difficulty, there is the *problem of the timing of the duplication*. In the all-important greenhouse sequence, the organisms that explode out of their shells are not wearing clothing. The pods produce naked duplicates who must acquire clothing, presumably from the people they duplicate. How, then, does the clothing worn by Becky in the tunnel appear on the pod after it has duplicated Becky's body? Are we to understand that in the sixty seconds between the time we see Becky splash water on her face and the time Miles returns to the tunnel and calls out to her, someone has managed to slip past Miles, enter the tunnel, and place a pod in Becky's proximity, which then duplicated her body, absorbed her mind the instant she fell asleep, put on her clothing, and disposed of her corpse? In addition to being inherently implausible, this is inconsistent with what we are shown about the duplication process, which is that it is not instantaneous.

(4) Finally, there is the *problem of how the pods are matched up with the persons (whose bodies) they duplicate.* There are several pods: a fully developed pod of Becky in the basement of her father's house, a developing one in Miles's greenhouse, and two undeveloped seed pods in the trunk of Miles's car as he and Becky try to make their getaway (the latter are set on fire when Miles discovers they have been put there by Mac [Dabbs Greer], the gas station attendant). Two seed pods, one for Miles and one for Becky, are placed in Miles's office when they are surrounded by Jack, Dan Kaufman, and Sam Janzek—once the three have become pod people. But which pod will grow into Becky's duplicate and absorb her mind while she sleeps? Presumably the one in closest proximity to Becky when she falls asleep, for if we say that all of them will become duplicates of Becky, then, per our first problem, Santa Mira will have numerous duplicates of Becky. But if we say only one of them, then do the other pods stop the duplicating process, develop into the duplicate of another person, remain dormant, decompose, or do something else?

Moreover, Miles sees a fully formed pod of Becky in the basement of her father's house, and yet the real Becky—the one Miles rescues—is upstairs asleep. Why has the changeover from person to pod not occurred? If the duplication can occur within sixty seconds (as in the tunnel sequence), why does the changeover from person to pod not occur at Becky's father's house when she is sound asleep (and remains so even as Miles lifts her up out of bed and brings her downstairs)? As we saw above, the tunnel sequence suggests that the changeover happens instantaneously, once Becky has fallen asleep, but the rescue sequence shows that it does not.

An Inference to the Best Explanation

The question for Drs. Bassett and Hill, as well as for first-time viewers of *Invasion of the Body Snatchers,* is, Why is Dr. Miles Bennell saying these extraordinary things about what is allegedly happening in Santa Mira? What is needed here is an account that provides the best explanation of Miles's actions. I take this to be equivalent to providing an interpretation of the film that provides the best explanation of Miles's behavior. And I shall take this to be equivalent to providing the best interpretation of the film itself (even if these are not quite the same thing).[30] Here, my reading of *Invasion* departs dramatically from previously offered interpretations. For my answer is that the best explanation of Miles's behavior is that Miles's story really is one of derangement and paranoia and that something has happened to Miles to induce an acute delusional psychosis.[31]

I can imagine someone asking, "But what about the pods themselves? We *saw* those pods, large as life, on the screen." But the viewer "sees" the pods only in the context of Miles's flashback narration. And his construal of the mechanisms of duplication, I have argued above, is inflected with incoherence, which is understandable if it is the product of a delusional psychosis. What Miles experiences as the "invasion" of the body snatchers is less an invasion of extraterrestrials or even "an epidemic mass hysteria," in the words of Dr. Kaufman, than a projection of Miles's own paranoia. *Invasion of the Body Snatchers* internalizes Miles's paranoia by having him frequently advert to his fear in his voice-over narration. It externalizes his paranoia with imagery such as dark, suffocatingly small closets, narrow hallways, tunnels, and other enclosures. Believing that he is being pursued by a nameless, unrelenting peril that threatens his very identity and existence as a human being, Miles lives in paranoid fear. During Miles's escape from the mob of pod people, Siegel and Mainwaring put him on a busy highway at night, and the cars, trucks, headlights, horns, and angry motorists shouting insults become an objective correlate of Miles's confused and disordered consciousness. The film is thus a devastating portrait of paranoia and a deeply subversive deconstruction of the conventions of 1950s science fiction film.

On the interpretation I am offering, the events we see on the screen between the opening and closing sequences in the hospital emergency room are products of Miles's psychosis and not those actually occurring in the life of *Invasion*'s leading character.[32] When Dr. Bassett asks, "Will psychiatry help?" Dr. Hill's answer sums it up: "If all this is a nightmare, yes." From the point of view of Dr. Hill, Miles's representation of events is easy to dismiss as delusional because fabrication, falsehood, and paranoid fantasy cannot be ruled out when the events he is recounting are so far removed from ordinary experience. As the eighteenth-century philosopher David Hume pointed out, a rational person proportions his belief to the evidence. Miles's claims require extraordinarily strong supporting evidence, which he does not provide to the doctors.

Seen in this light, the frame story is not adventitious, something tacked on as an afterthought. It is an essential element of the film. Contrary to the claims of its detractors, who invoke it to make Don Siegel the hero of a battle against studio philistines, the frame device raises the film to the ranks of a science fiction classic. By making the narration a flashback by Miles from the hospital, the prologue and epilogue give the first-person narration an ironic tone and help to explain why we encounter the four philosophical problems I identified above. As a product of paranoid delusion, Miles's account is sufficiently detached from reality to give rise to those inflections of incoherence that the four problems expose. Lacking a sufficient grounding in reality, unanswerable metaphysical questions about Miles's account of the mechanisms of duplication are bound to arise.

Picturing Paranoia

It might be thought that the ending of the film is inexplicable on my interpretation since, by hypothesis, the events that close out the epilogue are not delusions. During the epilogue, Dr. Hill comes to believe Miles largely because of an amazing coincidence. After Miles has concluded his story, two ambulance drivers wheel in an injured man on a hospital table. One of the drivers begins to chat with Dr. Bassett: "We had to dig him out from the most peculiar things I ever saw," he tells the doctor. Hearing this, Dr. Hill asks, "*What* things?" The driver says, "I never saw them before. They look like . . . great big *seed pods.*" Dr. Hill: "Where was the truck coming from?" Ambulance driver: "Santa Mira." On this basis, Dr. Hill orders the policemen to sound an all-points alarm, block all highways, stop all traffic, and call in every law enforcement agency in the state. He gets on the phone and tells the operator to get him the FBI. Dr. Hill takes these actions on the basis of the testimony of the ambulance driver, which suggests to him that Miles is, after all, of sound mind and that the story he has just told is true. But this is comparable to concluding that a story from an alleged eyewitness about a crash landing of extraterrestrials in Roswell, New Mexico, is true because one peripheral element of the story has been confirmed—for example, that bits of material with unusual markings on them were found at the site and confiscated by the air force.

The interpretation offered here makes the conclusion of *Invasion* even more devastating than the one found in the original version of Mainwaring's screenplay. For it now appears that even highly trained, presumably rational men like Dr. Hill can be led to draw unwarranted conclusions, and take far-reaching actions, on the basis of some weak confirming evidence. Dr. Hill's willingness to believe Miles's wild story on such slender evidence clearly violates Hume's idea that rationality requires us to proportion our beliefs to the evidence. The surprise finale, in which Dr. Hill accepts Miles's paranoid story, is an ironic reversal typical of film noir. The ending is thus to be interpreted as the filmmakers' use of irony to reinforce their own corrosive perspective, for the epilogue emphasizes the incongruities and illogic of Miles's paranoia and the absurdity of the noir predicament itself. If we take the epilogue at face value and not as an ironic comment on the proceedings, we have to ask on what authority Dr. Hill gives his orders to the policemen and why they carry them out without question. These features of the film should not, in my view, be dismissed as aesthetic flaws, as if the scene above could not mean more than meets the eye. My account proceeds under a principle of charity and assumes that the filmmakers were neither stupid nor intent on sabotaging their own film. It thus has the advantage of giving the filmmakers the benefit of the doubt. But whether this was their intention or not, the reading I propose makes the film less predictable, less hokey, and more interesting—in short, a better film.

Invasion of the Body Snatchers has been a catch basin for some of the most extreme currents of ideologically motivated film commentary. If one starts out with the intent to convey the idea that Siegel is misogynistic and reactionary, one must do so at the cost of overlooking the philosophical and aesthetic texture of this film. The misogyny and extreme individualist politics, if indeed they are there, can be used by the critic to camouflage the hatchet job that is being done. But it is a terrible waste of zeal. I have tried to provide a nonideological interpretation by taking a closer look at its key philosophical contribution. Throughout the film we are reminded of, in Miles's words, "what happens to people when they lose their humanity," as their specialness as persons recedes. They become pods: emotionless, affectless duplicates answerable to no higher purpose than the inexorable will to survive. Siegel and Mainwaring were quite likely attempting to say something on this order in *Invasion of the Body Snatchers*. Whereas Biskind interprets the film as a Cold War allegory and Steffen-Fluhr interprets it allegorically so that it symbolizes a man recoiling in terror from the vulnerability and commitment that love requires, for which he feels totally unprepared, I see it as an example of the stylized psychological realism of film noir.[33] Mainwaring was the author of the novel *Build My Gallows High,* from which he wrote the screenplay for the film noir classic *Out of the Past* (Jacques Tourneur, 1947). It is the story of an ill-fated noir protagonist who comes to grief at the hands of a femme fatale, told largely in flashback. If I am correct that *Invasion of the Body Snatchers* is a film noir, then it is not far-fetched to suggest that the film is really about the philosophical significance of paranoia. If *Invasion of the Body Snatchers* is viewed for its philosophical message rather than its midcentury politics, its message is that people are imperiled not only physically but also emotionally and, quite possibly, morally. They are in need of redemption or, in the resolutely secular world of *Invasion of the Body Snatchers*, rescue. Such rescue must come from within, as it were, by our own strenuous efforts to avoid succumbing to the receding humanity to which Miles alludes with simple eloquence in his conversation with Becky in his office as they await Jack's arrival. As in film noir, the stylized psychological realism that is so much a part of *Invasion* finds expression in a particularly complex form if the bulk of the film is a projection of Miles's paranoid fantasy, providing a self-referential dimension that adds to the film's philosophical cachet.

As we have seen, the interpretations of Biskind and Steffen-Fluhr do not make *Invasion* a very interesting picture. This is not only because they do not tell us much about the film that we could not see for ourselves, and because they rely on factual errors and dubious inferences, but also because they so shape their interpretations that *Invasion* is completely of its time and incapable of transcending that time. Under their controlling assumptions, the film does not live for us because it is merely a relic of its era's political assumptions and gender biases. They make the film seem too reductive, as if it were no more than a reflection of the social and political culture from which it sprang. There is no denying that *Invasion* bears the mark of its time, with all the limitations that that implies. When the students in Steffen-Fluhr's class viewed it in the early 1980s, they saw themselves as so remote from its 1950s atmospherics that its message was lost on them. But its influence is undeniable, and together with a handful of other science fiction films, it has achieved iconic status. *Invasion* is best remembered for its philosophical outlook, one that continues to persuade viewers to reflect on how, in Miles's words, "people have allowed their humanity to drain away. . . . We harden our hearts . . . grow callous. . . . Only when we have to fight to stay human do we recognize how precious it is to us." *Invasion of the Body Snatchers* attains philosophical stature because it readily permits application to our own time by tapping into this general truth about the human condition.

Notes

The line from *King Lear* that forms my epigraph is found in act 3, scene 4, line 21. I am grateful to Christeen Clemens for discussions of the film and for her many helpful comments on earlier drafts of this essay.

1. I discuss the nature, style, and themes of noir in both film and television in "An Introduction to the Philosophy of TV Noir," *The Philosophy of TV Noir*, ed. Steven M. Sanders and Aeon J. Skoble (Lexington: University Press of Kentucky, forthcoming).

2. The two interpretations appear in the Rutgers Films in Print series volume *Invasion of the Body Snatchers*, ed. Al LaValley (New Brunswick, NJ: Rutgers University Press, 1989). This showcase gives them a prominence that calls for philosophical engagement.

3. Don Siegel, interview by Guy Braucourt, in LaValley, *Invasion*, 159.

4. Don Siegel, interview by Stuart M. Kaminsky, "Don Siegel on the Pod Society," in LaValley, *Invasion*, 154.

5. According to Al LaValley, the first installment of the serial appeared in the November 26, 1954, issue. See LaValley, *Invasion*, 3.

6. Ibid., 4–6.

7. Siegel, interview by Kaminsky, 153.

8. Woody Haut, *Heartbreak and Vine: The Fate of Hardboiled Writers in Hollywood* (London: Serpent's Tail, 2002), 184.

9. Irving Singer's term describes the realistic detail with which Alfred Hitchcock photographed reality. Irving Singer, *Three Philosophical Filmmakers* (Cambridge, MA: MIT Press, 2004), 56.

10. This and subsequent dialogue from the film is taken from the continuity script prepared by Al LaValley for his *Invasion of the Body Snatchers*.

11. Continuity script, 31–109, 75.

12. Peter Biskind, *Seeing Is Believing* (New York: Pantheon, 1983), 111.

13. Michael Paul Rogin, "Kiss Me Deadly: Communism, Motherhood, and Cold War Movies," in LaValley, *Invasion*, 204.

14. Peter Biskind, "Pods, Blobs, and Ideology in American Films of the Fifties," in LaValley, *Invasion*, 194.

15. Biskind, "Pods," 193; Biskind, *Seeing Is Believing*, 138.

16. Continuity script, 51.

17. Biskind, *Seeing Is Believing*, 138.

18. Ibid.

19. Nancy Steffen-Fluhr, "Women and the Inner Game of Don Siegel's *Invasion of the Body Snatchers*," in LaValley, *Invasion*, 206–21.

20. Ibid., 207.

21. Ibid., 220, 219.

22. Ibid., 208.

23. Continuity script, 39.

24. Ibid., 65.

25. Steffen-Fluhr, "Women," 208.

26. Ibid., 220.

27. Ibid., 210.

28. Jack Finney, *The Body Snatchers* (New York: Dell, 1961), 90.

29. Ibid., 169.

30. My present purpose is primarily to sketch what I take to be the best interpretation of the film. Unfortunately, I cannot attempt a defense of my underlying assumptions about interpretation in general. For an approach to interpretation that has helped shape my outlook, see Richard A. Gilmore, *Doing Philosophy at the Movies* (Albany: State University of New York Press, 2005), esp. 154–62.

31. My favorite scenario is that he is undergoing a delayed reaction to a dose of slow-acting LSD slipped into a drink while he was at the medical convention from which he returns at the opening of the film, in the manner of *D.O.A.* (Rudolph Maté, 1950), whose protagonist dies from a highball laced with luminous poison. This is, of course, only one of many possible ways of accounting for his derangement.

32. The events fabricated by Miles, which are not part of the "real" world of the film but are his own fantasies and hallucinations, are metadiegetic material, as contrasted with diegetic material, which is part of the world of the film's characters, and nondiegetic material, which is not part of the world of the film at all. For example, the police bulletin transmitted over the radio is diegetic sound, whereas Carmen Dragon's obtrusive soundtrack is nondiegetic sound.

33. I am indebted to Jason Holt's discussion of film noir as "stylized crime realism" in "A Darker Shade: Realism in Neo-Noir," in *The Philosophy of Film Noir*, ed. Mark T. Conard (Lexington: University Press of Kentucky, 2006), 23–40.

THE EXISTENTIAL *FRANKENSTEIN*

Jennifer L. McMahon

In this essay, I shall offer an existential analysis of the science fiction classic *Frankenstein*. I shall argue that *Frankenstein* illustrates not only the anxiety that individuals have about death but also their tendency to deny it and their powerful desire to conquer it. Importantly, I shall also argue that *Frankenstein* illustrates the undesirability of death's defeat.

Though Mary Shelley's Frankenstein story has taken a variety of forms since it was published in 1818, certain elements of the story remain constant. Whether set in a gothic context or a modern lab, whether drama or comedy, the Frankenstein story examines the human desire not only to control nature but also to dominate it. It also examines the role technology plays in the achievement of human desire and considers the problems such technology may generate, ideas that are central to the science fiction genre. It tells the gripping story of Dr. Frankenstein, who, in his efforts to divine the "mysteries of life and death"[1] and determine to what extent the principles of life can be controlled, brings to life a monster that ultimately destroys (or attempts to destroy) all that Frankenstein loves. Since they clearly embody the classic traits of the science fiction genre and history has recognized them as classics in their own right, James Whale's cinematic rendition of the Frankenstein story (1931) and Mary Shelley's original text are the exclusive foci of my analysis. Before turning to the Frankenstein story and discussing what it has to say about death, I will establish a theoretical framework for this discussion to introduce readers to the existential concepts that will be central to the analysis of the film.

The Devil Within: Heidegger on Human Mortality

Though existentialist philosophers are known for their preoccupation with morbid topics, few of them place as much of an emphasis on death as Martin

Heidegger. In *Being and Time,* Heidegger examines the nature of human being, what he terms *Dasein.*[2] He asserts that there are a variety of qualities that characterize human beings, and he enumerates them in the text. The quality that receives the most attention from Heidegger, and the one that is arguably the most important to him, is "being-toward-death." Heidegger emphasizes that humans, in addition to being "being[s]-in-the-world," social beings, and "beings-toward-possibility," are beings who are mortal. Indeed, Heidegger focuses on finitude in his discussion because he contends that it is primarily the anxiety that humans have over their mortality that leads most of them to live "inauthentic[ally],"[3] or in denial with respect to the nature of the human condition. Though Heidegger did not see overcoming death as a possibility, he would assert that it would remove one of the most formative aspects of our being. As such, it would change who we are. Though becoming immortal may well be one of our most deep-seated desires, *Frankenstein* illustrates that actual mastery of death could well result in a loss of humanity. It shows that victory over mortality could make us monsters.

In *Being and Time,* Heidegger analyzes the nature of human beings. Heidegger concludes from his analysis that humans are essentially temporal beings who constantly change, whose existence depends upon both the presence of a particular type of environment and interaction with others, and who display a fundamental "concern"[4] about being. Though Heidegger attends carefully to each of these aspects of the human condition, death is his focal point because it so shapes human behavior and experience. He foregrounds death because it is the principal cause of human concern.

A recurrent claim in *Being and Time* is that humans are characterized by concern or "care" about being. Though Heidegger maintains that human concern can and should extend beyond the individual, he notes that the being about which each one of us is fundamentally concerned is "always my own." As Heidegger states, "Dasein is a being which, as a being-in-the-world, is concerned [primarily] about itself."[5] Though other things can motivate care, Heidegger asserts, the primary reason humans are concerned about their being is because it is finite.

Of course, humans are not the only beings that die. However, humans are unique in our awareness of death. As Heidegger and others attest, the presence of reflective consciousness in humans makes it possible for us to have an awareness of both our being and our imminent not being, forms of awareness that other animals lack. Celebrated for centuries as the feature that distinguishes humans from other species, reflective consciousness affords humans the opportunity to ascend to spiritual heights. However, it also makes humans acutely aware that we are physical beings vulnerable to injury and death. In Whale's version of *Frankenstein,* we see this awareness of death—and the pain that accompanies it—emerge in the monster when he inadvertently kills the peasant girl by the lake. According to Heidegger, human nature is characterized by concern because unlike other animals, humans are conscious of the fundamental tenuousness of their being. We are anxious because we know we exist in a world that is both something on which our existence depends and "something by which [we] can be threatened." We exhibit concern because at the deepest levels of our being we know we are finite, that "death is a way to be that [we take over] as soon as [we exist]."[6]

While Heidegger contends that human concern is rooted primarily in our awareness of death, he does not believe that most people formally acknowledge their mortality. Instead, he asserts that most people deny the reality of their death, just as they do most other aspects of their being. Heidegger uses the term *inauthenticity* to describe this state of denial. People who are inauthentic deny the true nature of the human condition, whereas those who are authentic do not. Most people are inauthentic because they are trying to avoid anxiety.

According to Heidegger, all humans experience anxiety or angst. Indeed, Heidegger describes anxiety as the "basic mood" (i.e., affective state or disposition) of Dasein. For him, anxiety is an unavoidable consequence of being. He asserts, "That about which angst is anxious is [its] being-in-the-world," particularly its "death." Ultimately, awareness of the nature of one's being, specifically awareness of one's mortality, incites anxiety. Because most people find feelings of anxiety unpleasant, they seek to escape these feelings by "cover[ing] over" or denying that which stimulates them. According to Heidegger, no person is truly ignorant of the nature of his or her being. Anxiety represents the individual's original and visceral understanding of his or her being. In Heidegger's estimation, our gut knows the truth of our being before our mind ever grasps it. Indeed, Heidegger asserts that authentic awareness of being (i.e., cognitive apprehension of the true nature of one's being, accompanied with emotional acceptance of it) results from reflecting upon the affective responses we have to existence. However, he contends that most of us resist doing this (and instead fall prey to inauthenticity) because of the discomfort anxiety creates. Thus, rather than acknowledge the existential truths to which feelings of anxiety attest, most of us are inauthentic and engage in what Heidegger describes as an "evasive turning away" from the aspects of our being we find mentally and emotionally too "burden[some]"[7] to bear. To the extent that death not only affects one's being (e.g., through the loss of loved ones) but also marks the end of it, the prospect of death motivates more anxiety than any other aspect of the human condition and is therefore the most frequently denied.

The Denial of Death: Essential Illusion or Fatal Flaw?

According to Heidegger and the contemporary psychological theorist Ernest Becker, the denial of death is an extraordinarily pervasive, if not universal, phenomenon. Its ubiquity is a function of the fact that "the idea of death, the fear of it, haunts the human animal like nothing else."[8] As Heidegger and Becker explain, the denial of death takes a variety of forms and has specific effects. Most frequently, individuals deny death by preoccupying themselves with the mundane affairs of life. Heidegger and Becker agree that while most of us readily admit the mortality of humans generally, few accept this general fact as a personal certainty.[9] Instead, we avoid making the terrible admission of our mortality by throwing ourselves into "the manageable urgencies and possibilities of the everyday," most frequently into a life of uninhibited "busyness."[10] To the extent that most others are engaged in a similar enterprise, culture comes to both reflect and facilitate the activity of denial: it celebrates youth, discourages discussion of mortality, and promotes the incarceration in hospitals and nursing homes of individuals who might catalyze awareness of our mortality. We avoid death whenever possible, and when we must confront it, we often dress it up to disguise its finality.[11]

Though simple avoidance of thought about (and proximity to) death is the most common form of its denial, Heidegger and Becker both note that a preoccupation with death, though unusual, can also serve this purpose. In *The Denial of Death*, Becker examines this fascination with death, referring to it as the "demonic extreme" of "defiant Promethianism."[12] Rather than using the conventional technique of repressing anxiety about death by avoiding thought about it, the individual who is fascinated with death attempts to inoculate himself against anxiety by conjuring the hope that death can be controlled, even conquered. As Becker states, though this approach is uncommon, this individual's objective is analogous to that of the more conventional type. Specifically, his goal is still "to deny his lack of control over events, his powerlessness, his vagueness as a person in a mechanical world spinning into decay and death."[13] Simply put, the Promethean type's objective is still the denial of death. He simply takes the offensive approach as opposed to the defensive.

The question remains whether the denial of death is a "vital lie"[14] (i.e., a productive coping mechanism) or a fatal flaw. While Heidegger and Becker understand the inclination to deny death, neither supports the practice. Whether denial takes the conventional form of evasion or is expressed in an unusual fascination with death, they agree, it is unproductive. Though Heidegger denies that he makes any normative claims in *Being and Time,* it is impossible to interpret his discussion of authentic being-toward-death as anything but prescriptive. Whereas he asserts that denying death (i.e., inauthentic being-toward-death) results in a "leveling down of possibilities"[15] for the individual and "entanglement" in the "they," he claims that authentic being-toward-death (i.e., forthright acceptance of death) restores one's individuality, heightens one's passion, evokes both a sense of concern for and responsibility to others, and liberates one from "one's lostness in chance."[16]

Becker is not as reserved as Heidegger with respect to offering evaluative judgments. He states explicitly that people should overcome their tendency to deny death. While he acknowledges that it is impossible to confront one's mortality "without anxiety," he asserts that it is nonetheless imperative that we acknowledge and accept our finitude. In his estimation, not only will this recognition create an opportunity for humans to exhibit "new forms of courage and endurance" in confronting the stimulus of their greatest "terror," it will also preclude them from engaging in commonplace but ill-advised and deeply ironic "protests against . . . natural reality [and] the human condition."[17] As Becker explains, though it placates anxiety, denying death has a number of negative effects.

The first problem with the denial of death is that it never achieves its ultimate goal: the eradication of anxiety. Like Heidegger, Becker believes that anxiety is an inescapable feature of our being. Though techniques of denial can succeed in repressing anxiety, they cannot eliminate it. Instead, when denied forthright expression, our anxiety about death transforms into other fears (e.g., fear of heights or spiders). Because our concern about death is never truly eradicated, denial is an ongoing project that implies "a constant psychological effort to keep a lid on [an undesirable truth] and never inwardly relax our watchfulness."[18] Because of the expenditure of energy it requires and the evasions it entails, denying death proves an exhausting enterprise that eats up our lives without ever achieving its end. Whereas authentic being-toward-death allows individuals to honestly confront the object of their anxiety, denial condemns individuals to a life on the run from an assailant they cannot see.

The second negative effect of the denial of death is that it derides our social relations by discouraging individuals from communicating regarding matters of utmost concern. Although frank communication about death (and the understandable anxieties we have about it) would foster the development of empathy, intimacy, and solidarity among individuals, cultural prohibitions against such discussions limit interpersonal communication to the "tranquillizing [but] alienating"[19] conversations about superficialities that dominate much public intercourse.

Finally, the denial of death encourages an unconscious antagonism toward the physical body and the natural world. Because the empirical world is the context in which death occurs and the material body is the victim of this offense, people's natural anxiety toward death (which Becker argues is an expression of one's instinct for self-preservation) often manifests itself as antagonism (often unconscious) toward the natural world and the physical body. It is ultimately both exceedingly unwise and ironic, though understandable, for a being who is dependent upon the natural environment and defined by her embodiment to hold attitudes or engage in practices that are antagonistic to nature or the human form. For Becker, such actions are born of blindness to the human condition, blindness that is a consequence of denial. Rather than persist in denial, Becker states, "I think that taking life seriously means . . . that whatever man does on this planet has to be in the lived truth of the terror of creation, of the grotesque, of the rumble of panic underneath everything. Otherwise it is false."[20]

Victor Frankenstein and His Defeat of Death: Victory at What Price?

Frankenstein illustrates the anxiety that individuals have about death, their tendency to deny it, and their desire to conquer it. Though it focuses most explicitly on the desire that humans have to dominate death, Whale's cinematic version of Shelley's original text foregrounds the anxiety that individuals have about death in its opening scene. The scene depicts a burial. Rather than focus on the coffin that is being interred, Whale instead attends to aspects of the setting and to the affective responses of the onlookers. Like the body being laid to rest, the scene is lifeless. Without a blade of grass or a single leafed tree, the setting is barren and wasted. Skewed angles predominate and generate unease by disturbing one's perception of the horizon. Like the setting, the attendants at the burial are disturbed. Overwrought with grief, they emit sobs that dominate the opening sound track. The plaintive cries of the attendants are punctuated and intensified by the ominous tolling of a bell.

Later in Whale's version, we see anxiety about death manifested again, this time in someone whom we might not expect it to affect: Fritz (Dwight Frye), Frankenstein's deformed and demented assistant. Though Fritz's (and Frankenstein's) glee contrasts sharply with the despair shown by the attendants in the opening scene, Fritz clearly exhibits anxiety toward death in the crucial scene where he takes the wrong brain. In this scene, Fritz enters the lecture hall where Dr. Waldman (Edward Van Sloan) has just concluded a discussion of what distinguishes a normal from an abnormal brain. With the lights of the lecture hall extinguished, Fritz approaches the podium in order to procure the normal brain for Frankenstein's use. The cargo safely in his arms, Fritz attempts to depart the darkened room but is startled by the shadow cast by the model skeleton that stands at the front of the podium. This shadow of death dominates the frame. It looms over the cowering Fritz, who, in his panic, fatefully drops the normal brain.

Ultimately, Whale's decision to allude to humans' anxiety about death not only helps establish central elements in his plot but is consistent with the original Frankenstein story. In Shelley's novel, anxiety over death is presented as the underlying motivation for Victor Frankenstein's pursuits. Central to the psyche of Shelley's "mad scientist" is the death of his mother while he is still a boy. This loss frames death as fundamentally "evil,"[21] a "spoiler," an "enem[y]" that robs Frankenstein of his primary source of love and security. Though he later comes to see death in a different light, Victor Frankenstein initially expresses only "deep and bitter agony"[22] at the prospect of death. He exhibits the anxiety about death that Heidegger and Becker describe.

In addition to illustrating the anxiety that humans feel about death, *Frankenstein* also illustrates their tendency to deny it. In Whale's version, this denial is exhibited in Frankenstein's repeated references to corpses as not dead but resting. As Frankenstein (Colin Clive) says to Fritz when they obtain the body buried in the opening scene, the figure is not dead, but instead "he's just resting, waiting for a new life to come." Similarly, when Frankenstein invites Dr. Waldman, Elizabeth (Mae Clarke), and Victor Moritz (John Boles) to witness the quickening of his monster (Boris Karloff), Victor denies that he is raising the dead. He contends that because he has made his monster only from pieces of corpses that it is not itself dead but a new creation that is waiting to live. Though Frankenstein is convinced by this logic, neither his witnesses nor the audience finds it especially compelling. Instead, it clearly seems to be expressive of psychological denial.

The denial of death is also apparent in Shelley's original text. It is particularly evident in Frankenstein's refusal to regard death as an absolute. Instead, he states, "Life and death appeared to me [as] ideal bounds, which I should . . . break through . . . [so] I might in the process of time . . . renew life where death had apparently devoted the body to corruption."[23] Here, it is evident that, rather than regarding death as a permanent state, as an inalterable ontological fact, Frankenstein sees it as a conquerable foe whose defeat is not only possible but also desirable.

In addition, in both Whale's version of the Frankenstein story and Shelley's original text, the tendency to deny death is made evident in a more subtle, symbolic fashion. Specifically, in both accounts, the horror and antagonism expressed toward the monster and its expulsion from society are symbolic both of the aversion we have toward death and our tendency to deny it. Ultimately, though Frankenstein's monster serves other purposes, one of the creature's primary functions is to personify death. This personification is most evident in Shelley's novel where the monster is repeatedly described, like death so often is, as "horrid" and "unearthly," a "fiend who snatche[s] . . . every hope of future happiness."[24]

The identification of the monster with death is made even clearer by the fact that in Shelley's novel the creature kills everyone Frankenstein loves. Thus, through his murder of William, Elizabeth, Henry, and others, the creature not only exacts his revenge upon his creator but effectively becomes the grim reaper. Because of the role that the monster plays, the terror and animosity displayed toward him symbolize the fear and aversion that humans have toward death. Likewise, the casting out of the monster, his shunning by society, symbolizes the tendency that individuals have to deny death, to exorcise it from their ordinary occupations, to put it off to a safe (and forgettable) distance.

Object Lessons in the Danger of Denial

As both Whale's version and Shelley's original text make clear, the type of denial in which Frankenstein is engaged is not the conventional avoidance of death that most of us employ. Rather than trying to avoid death, Frankenstein exhibits the unusual fascination with death that Becker describes as "defiant Promethianism." Indeed, Shelley subtitled her work *The Modern Prometheus*. As discussed previously, characteristic of this form of denial are the desire to conquer death and the belief that such a victory is possible. Typically, individuals exhibiting this form of denial also manifest a desire to achieve transcendence over conventional human limits; specifically, they seek to overcome the ontological constraints that simultaneously define humanity and generate anxiety.

Whale's Frankenstein clearly exhibits the desire to conquer death. Unaffected by his former professor's warning to avoid "dangerous" and potentially "deprav[ed]" experiments, Frankenstein responds, "Haven't you ever wanted to do anything dangerous? . . . Where would we be if nobody tried to find out what lies beyond? You never wanted to look beyond? To know . . . what changes darkness into light? What eternity is, for example." Subsequent to the animation of his monster, Frankenstein not only offers the film's most famous line—"It's alive. It's alive"—but also revels with maniacal glee in the inhuman power he has exercised, shouting, "Now I know how it feels to be God."

In the original text, Shelley also attends to the desire to control nature, specifically the desire to conquer death. Conveying his story to Robert Walton, Shelley's Frankenstein confesses that he was driven to his research by a "fervent longing to penetrate the secrets of nature." He states unequivocally, "I entered with the greatest diligence into the search of the philosopher's stone and the elixir of life." He admits that his goal is to achieve "immortality," not just for himself but also for others. In part motivated by egoism and the "glory that would attend [such a] discovery," Frankenstein seeks to "pioneer a new way," one that defies rather than accepts death as destiny. He states, "[My goal is] dominion . . . over the elemental foes of our race"; it is "to banish disease from the human frame and render man invulnerable to any but a violent death."[25]

Ultimately, in both Whale's version and Shelley's original text, the character of Frankenstein not only epitomizes the desire to conquer death but also sates this desire when he brings his monster to life. While overcoming death could be heralded as the greatest victory humans could achieve (to the extent that it would remove one, if not the principal, cause of human anxiety), existentialists like Heidegger suggest that our humanness is bound up with our mortality. If finitude does shape our humanness, both the effort to overcome and actual mastery of death could compromise our human sensibility. Intended to serve not only as an entertaining piece of fiction but also as a cautionary tale, *Frankenstein* articulates precisely this concern.

Though the normative function of Shelley's narrative is more obvious than that of Whale's, both Whale and Shelley offer object lessons in the danger that accompanies efforts to control nature and redefine human potentialities and powers. They do so by illustrating the negative consequences that follow from Frankenstein's uncritical pursuit of power and mastery of "the principle of life."[26]

The first consequence that warrants consideration is the effect that Frankenstein's activities have on his own person. In Whale's version, Elizabeth and Dr. Waldman attest that Frankenstein has been "greatly changed" by his research. Frankenstein's friend Victor Moritz says he has become "strange." Not surprisingly, Whale makes the change in Frankenstein's character evident most clearly through cinematic means. Colin Clive's maniacal looks, charged dialogue, and paranoid concern create an impression that his experiments have brought him to the verge of losing not only his sanity but also his humanity.

In Shelley's original text, Frankenstein himself testifies to the negative effects that his efforts to overcome death have on his character. The changes he describes are both physical and mental. Physically, Frankenstein asserts, his preoccupation with giving life to his creature becomes so overwhelming that he neither eats nor sleeps. Consequently, he becomes "emaciated," a mere "skeleton,"[27] a shadow of his former self, an approximation of death. His victory over death brings no improvement in his physical health. Instead, he deteriorates further. He is subject to recurrent illness, and others comment repeatedly upon his pathetic physical condition.

As significant as the physical changes that affect Frankenstein are, the psychological changes that take place during the course of his involvement in activities that he himself describes as "unhallowed" are even more noteworthy. Frankenstein states that during his trips to the charnel houses and the construction of his monster, he was like a man possessed: "I would often lose all self-command, all capacity of hiding the harrowing sensations that would possess me during the progress of my unearthly occupation." Throughout the novel, Frankenstein describes himself as being subject to a loss of reasoned perspective, a "fervour" so powerful that it renders him incapable of "consider[ing] the effect[s] of what [he] was . . . doing." Though he admits, upon reflection, that his "heart often sickened at the work of

[his] hands" and that he sensed his activities were "unwholesome" and "not befitting of the human mind," Frankenstein nonetheless throws caution to the wind and pursues his "dream"[28] of defeating death.

The physical and mental changes that affect Frankenstein are relevant in that both testify to the erosion of his humanity. In the case of the physical changes, Frankenstein's human substance is literally reduced by his efforts to control nature and overcome death. Though he strives to generate life, he comes to look like a skeleton, to symbolize death. Likewise, though Frankenstein "succeed[s] in discovering the cause of [the] generation of human life; nay more . . . [in] bestowing animation upon lifeless matter," he testifies that his reasoning skills, particularly his capacity to consider consequences, are compromised. Rather than determining his own actions through reason, Frankenstein becomes a "slave"[29] to his desire to eradicate death. Though it can be argued that Frankenstein's transformation results from the perversion of reason rather than the loss of it, the symbolic effect remains the same. To the extent that the presence of a particular type of reason has traditionally served to distinguish humans from other species, a reduction in this capacity, or the distortion or loss of it, is indicative of a loss of humanity.

In addition to eroding his humanity, the Promethean denial of death that Frankenstein hopes for, and ultimately achieves, also degrades his relationships with others. This is made clear in Whale's film and Shelley's original novel, both of which depict Frankenstein as estranged from others. While social taboos against the desecration of bodies explain Frankenstein's departure from lawful society in order to conduct his experiments, the estrangement is not purely practical. Instead, Frankenstein's estrangement from others alludes symbolically to his transformation into a being not bound by death, into someone who—like his monster—is not fully human. In Shelley's novel, Frankenstein attests that his success creates "an insurmountable barrier . . . between me and my fellow[s]." His relations to others are compromised because his desire to overcome death supersedes all other considerations. Though he claims that he wants to save humanity, he ultimately uses humans as a means to an end. He becomes so enthralled by his project that he not only becomes insensitive to other individuals but objectifies and even "tortures the living animal"[30] in pursuit of his goal. Both Whale and Shelley show the estrangement from others as negative (and degrading to one's humanity) through their contrast between the social and natural environments. For example, Whale confines Frankenstein to an abandoned mill amid inhospitable terrain and juxtaposes this barren, inhuman setting with shots of the lavish interiors of the Frankenstein mansion and the joyous festivities that take place in the town. Shelley not only establishes a contrast between the social and natural environments but makes the effect of this contrast clear by depicting nature as progressively more threatening to human life, culminating in Frankenstein's pursuit of his monster to the absolute and inhuman isolation of the frozen Arctic (a scene reminiscent of Dante's *Inferno*).

Heidegger would not find it surprising that Frankenstein's efforts to deny death have a negative effect on his relations with others. For Heidegger, when one denies death, one denies that one shares in the common destiny of humanity. In his view, forthright awareness and acceptance of death not only have the positive effect of heightening our appreciation of the limited time each one of us has but also serve as the basis for genuine community.[31] Despite our differences, humans are all (at least at present) in the same boat with respect to their mortality. In his view, authentic awareness of this common fate creates a psychological bond between individuals and fosters a sense of moral obligation. Simply put, it promotes empathy, social concern, and a sense of solidarity with others. Though Heidegger resists asserting that authenticity is preferable to inauthenticity, he does claim that people who live in denial of their mortality (i.e., who exhibit inauthentic being-toward-death) typically lack the "conscience"[32] of those who are authentic toward death. Presumably because they are preoccupied with the exigencies of everyday life, they are often not as sensitive (both to other people and to the environment) as they could be, are more prone to objectify others, and are more likely to limit themselves to superficial relationships. Although Heidegger did not see the overcoming of death as a possibility, it seems clear that this would likely exacerbate the negative effects he describes. The Frankenstein story suggests this is the case.

A final negative consequence to which *Frankenstein* alludes is that increased control over nature, specifically our mortality, will lead to the generation of intractable moral dilemmas. Both Whale and Shelley articulate this concern. In both of their works, Frankenstein is faced with a dilemma: should he preserve his creation or should he destroy it? Though the renditions differ somewhat, both Whale and Shelley examine the moral dilemmas that follow from seizing control of nature, particularly taking control of life and death. For example, though the realization comes late, Shelley's Frankenstein recognizes that his uncritical desire to control life and eradicate death may have unwittingly placed him on a slippery slope. Having created a being, he feels some obligation to it. However, when the monster commands him to fashion it a mate, Frankenstein realizes that fulfilling this command might threaten humanity. In short, he senses a conflict between the obligation he has to his creation and the obligation he has to humanity. Frankenstein concludes that his duty to humanity is "greater because [it] include[s] a greater proportion of happiness and misery." Frankenstein recognizes that he has engaged in "unhallowed arts"[33] and ventured into territory where even angels should fear to tread.

Grave Concerns: Is Overcoming Death Possible and Desirable?

While some might argue that the foregoing discussion is irrelevant because we do not seem anywhere close to the defeat of death depicted in *Franken-stein*, the topic warrants consideration to the extent that disdain for, and a desire to control, death is alive and well in modern culture. Not only do our everyday language and practices deny death, but many of the religious and philosophical systems to which we subscribe do so as well.[34] The denial of death is also readily apparent in the medical context, a context where death is already being forestalled and research is being conducted whose goal is the indefinite extension of life.[35]

Interestingly, in Shelley's original text, Frankenstein actually revises his position on death. Rather than regarding death as the unequivocal enemy, he comes to recognize that for an individual whose life offers nothing but misery, death is likely something sought. He recognizes that his obsession with overcoming death has led him to disregard life, specifically the lives of his loved ones, those he endangered, and the being he created. He declares that he was mad and actively strives to preclude others from pursuing similar paths.

Frankenstein's story and his reform offer a message to a contemporary audience, who may experience some of the same negative effects that Frankenstein does if they persist in the denial of death. Through an extreme example, *Frankenstein* illustrates some of the serious problems that come from a refusal to accept mortality. It shows how the preoccupation with death's defeat can foster insensitivity to others and a disproportionate fascination with death rather than attention to the matter and meaning of life. Because the denial of death is a common phenomenon, these problems exist already and would only be augmented by the indefinite extension of life.

Though *Frankenstein* depicts the actual defeat of death, I hope to have shown that death need not be wholly conquered in order for these problems to arise. While the Promethean form of denial discloses the most obvious dangers of denial, conventional avoidance is equally problematic. As Heidegger and Becker explain, though it may be natural for us to deny death, it is not beneficial. Heidegger asserts that while the conventional denial of death is pervasive, "tempting and tranquillizing, it is at the same time alienating."[36] While the denial of death may bind us to others by virtue of a common project, it does not afford us the opportunity for true community with others or genuine discovery of ourselves. Instead, the denial of death promotes what Heidegger calls a life of "groundless floating." Rather than fostering genuine connection by allowing individuals to discuss serious existential matters, denial limits individuals to "idle talk" and to a life "of busyness . . . taking care of things." Because denial never actually eliminates anxiety about death, individuals who persist in the denial are both driven by and driven to distraction. As such, they slip easily into insensitivity. Moreover, as Heidegger states, this "[phases] out the possible" and ironically destines individuals to a sort of "nonbeing . . . in which [individuals] mostly maintain [themselves]." Whereas authentic being-toward-death presents the individual with the daunting task of "underst[anding] . . . and endur[ing death] as possibility," Heidegger maintains, it is through the confrontation with, and acceptance of, death that one discovers one's individuality. For Heidegger, until one recognizes one's being is time and one's time is finite, one is likely to waste both. One is also likely to be blind to one's connection (and obligation) to other people and the physical world. By foregrounding the problems that can issue from a denial of death, I have tried to encourage greater appreciation for the positive role mortality plays in forging our humanity, an effort that, if successful, might help people come to better terms with what Heidegger calls their "ownmost nonrelational possibility."[37] By fostering greater acceptance of death, I hope to promote fuller appreciation of life.

Notes

1. This and all other quoted dialogue, unless otherwise noted, is from *Frankenstein* (James Whale, 1931).

2. Martin Heidegger, *Being and Time* (Albany: State University of New York Press, 1996), 9.

3. Ibid., 233, 37, 40, 40.

4. Ibid., 114.

5. Ibid., 37, 40, 134.

6. Ibid., 129, 228.

7. Ibid., 172, 175, 232, 234, 128, 127.

8. Ernest Becker, *The Denial of Death* (New York: Free Press, 1973), ix.

9. Heidegger, *Being and Time*, 234.

10. Ibid., 239, 166.

11. See C. W. Wahl, "The Fear of Death," in *The Meaning of Death*, ed. Herman Feifel (New York: McGraw Hill, 1959).

12. Becker, *Denial of Death*, 85.

13. Ibid., 84.

14. Ibid., 51.

15. Heidegger, *Being and Time*, 181.

16. Ibid., 164, 243.

17. Becker, *Denial of Death*, 58, 279, 11, 33.

18. Ibid., 17.

19. Heidegger, *Being and Time*, 166.

20. Becker, *Denial of Death*, 238.

21. Mary Shelley, *Frankenstein* (Oxford: Oxford University Press, 1969), 43.

22. Ibid., 43, 100, 88.

23. Ibid., 54.

24. Ibid., 57, 219, 197.

25. Shelley, *Frankenstein*, 39, 40, 46, 40, 48, 28, 40.

26. Ibid., 51.

27. Ibid., 54, 183.

28. Ibid., 55, 152, 28, 165, 164, 56, 56, 58.

29. Ibid., 52, 153.

30. Ibid., 158, 54.

31. Admittedly, to the extent that most people seek to deny death, they are united in a common project. However, Heidegger denies that inauthenticity promotes community. Instead, he sees it as detrimental to it. Though the denial that is characteristic of the inauthentic life causes one to live like other people (e.g., one emulates the evasive maneuvers to which one is witness), it does not foster genuine connection to them. Thus inauthenticity generates the appearance, but not the actuality, of community.

32. Heidegger, *Being and Time*, 248.

33. Shelley, *Frankenstein*, 217, 89.

34. Wahl, "Fear of Death," 18. Here Wahl discusses how we frequently deny death by referring to it as "passing" and by engaging in practices like "prettify[ing] the corpse" to make it appear as if the individual who has died is sleeping. He states, "We flee from the eventual reality of our deaths with such purpose and persistence and we employ defenses so patently magical and regressive that these would be ludicrously obvious to us" were the activity of denial not so pervasive.

35. See, for example, Dr. Aubrey de Grey's Strategies for Engineered Negligible Senescence at http://www.sens.org/.

36. Heidegger, *Being and Time,* 166.

37. Ibid., 165, 181, 181, 164, 241, 237.

Part 2

EXTRATERRESTRIAL VISITATION, TIME TRAVEL, AND ARTIFICIAL INTELLIGENCE

TECHNOLOGY AND ETHICS IN
THE DAY THE EARTH STOOD STILL

Aeon J. Skoble

"Such power exists?"
　　　—Professor Barnhardt

Robert Wise's 1951 movie *The Day the Earth Stood Still* is generally regarded as a classic of science fiction film. At least as a working definition of the genre, I take science fiction to be that branch of literature (and by extension films) that deals with the effects of science or technology on the human condition or that explores the human condition via science. This can include utopian or dystopian future societies, of course, but *The Day the Earth Stood Still* is set in early 1950s America.

Science and technology play several roles in *The Day the Earth Stood Still*, most obviously in the presence of the spaceship that brings the alien Klaatu (Michael Rennie) and the robot Gort (Lock Martin) to Earth. But the other relevant technology is of human origin: the nuclear bomb and the nascent space program. Klaatu's mission to Earth, we learn at the film's climax, is to deliver a warning: as long as we fought among ourselves, this was not a matter of concern to the galactic civilization he represents. But now, we have nuclear weapons and are on the verge of space flight—this presents a threat to others and will not be tolerated. The film, then, both addresses the impact of the new human technology on society and forces us to consider the ethical ramifications of the alien technology. In this essay, I will argue that the two are related and that the latter is allegorical and thus has specific importance to actual ethics, especially regarding issues such as preemption and containment in war, paternalism, and self-defense.

Tellingly, *The Day the Earth Stood Still* has very few special effects and derives its power from the story and its characters. Perhaps for this reason some see the film as a Christian allegory: Klaatu comes from beyond the world, with a message of love and peace, is misunderstood and killed, gets resurrected, and returns to the sky. Despite his adopted identity as "Carpenter," I find this interpretation implausible, first because Klaatu's message is not that we must all love one another—it's OK if we don't, actually, as long as we don't threaten others—and second because Jesus didn't threaten to have his robot friend blow up the planet if we didn't listen.[1]

Klaatu's arrival in Washington DC is inauspicious from the start. His spaceship is met by hundreds of soldiers, complete with tanks and artillery. Crowds of people have gathered around also, so both fear and curiosity are on display. But as soon as the ship's door opens and Klaatu emerges, his space helmet obscuring the fact that he looks just like a human, curiosity gives way to fear: the civilians pull back, screaming, and the soldiers get nervous. Klaatu is literally instantiating the stereotypical "Take me to your leader" appearances of pulp science fiction aliens: he really does want to meet with the president and indeed has a nice gift for him. But a nervous and trigger-happy soldier, taking the gift for a weapon, shoots Klaatu. This causes Gort to melt all the soldiers' armaments, and he might well have started disintegrating the soldiers if Klaatu, not seriously injured, had not told him to stand down. Even this opening sequence suggests that technology (in this case a simple handgun) can be used ethically or unethically. Why did the soldier panic? Why did he assume the odd-shaped object in Klaatu's hand was a weapon? The soldier had no idea what was going on and hence shot an ambassador from another world and destroyed what would have been a nice gift for the president in the process.

It's a cliché in moral philosophy that *ought* implies *can*, but no one seriously defends the reverse. Just because we *can* kill the alien doesn't mean that we should. Indeed there are two reasons why we should not: first, it's only justifiable to kill an alien who is attacking you, not one who comes in friendship bearing a gift, and second, if Klaatu had been killed by the soldier, Gort would have killed all the soldiers and maybe even destroyed Earth. Thinking about these reasons why the soldier acted badly in shooting Klaatu thus points us toward more general ethical principles about the use of force. First of all, someone has to *do something* in order to deserve being shot. Just showing up to say hello isn't sufficient. Second of all, it's not prudent to attack someone whose retaliation will be devastating or whose retaliatory capabilities are unknown.

The irrational fear of the unknown plays a key role in the development of the story. Klaatu wants to speak not only to the president of the United States but to all world leaders. He's frustrated by the leaders' unwillingness to do so, which is partly motivated by fear of Klaatu, as well as by Cold War–inspired fear of each other. When Klaatu realizes that his (largely unnecessary) hospital stay has turned into (a futile attempt at) imprisonment, he leaves, hiding in plain sight as a resident of a boarding house while he tries to think of a way to get his message across. With the help of young Bobby Benson (Billy Gray), Klaatu starts to learn more about Earth, and he eventually decides to get his message across by addressing the scientific community, rather than waiting for the world's leaders to resolve their political stalemate. So he seeks out Professor Barnhardt.

Let Him Through, He's a Scientist

Professor Barnhardt (Sam Jaffe) is an archetypal 1950s movie scientist, with wild hair, eccentric mannerisms, a practical-minded housekeeper, and a sense that scientists could run things better than politicians if given the chance. ("We scientists are too often ignored, or misunderstood," he informs Klaatu.[2]) Although he insists that his full message be reserved for the international group of scholars the professor is assembling, Klaatu explains part of his mission to Professor Barnhardt: "So long as you were limited to fighting among yourselves with your primitive tanks and aircraft, we were unconcerned. But soon, one of your 'nations' will apply atomic energy to spaceships. That will create a threat to the peace and security of other planets. That, of course, we cannot tolerate. . . . I came here to warn you that by threatening danger, your planet faces danger."

Even this partial disclosure of Klaatu's mission points to interesting moral considerations. First of all, Klaatu and the people he represents are not paternalistic: they do not have an idealistic mission to stop all fighting.[3] They respect our autonomy in that regard. It might be regrettable, evidence of the primitive state of our society, that we fight among ourselves, but it's not seen as the business of Klaatu's society to interfere. Perhaps it's a phase that many civilizations go through, and we'll eventually grow past it. Perhaps Klaatu's society recognizes that it lacks the wisdom to know how to interfere in our local problems. In any event, Klaatu's position is that our autonomy regarding local disputes does not trump other planets' right to peace and security. So if we're going to "export" our problems into space, then it *is* the business of Klaatu's society to interfere. This is one way of extrapolating the familiar harm principle to the macro level: you may have the right to harm yourself, but you definitely lack the right to harm others.[4] Similarly, members of society A may have the right to fight among themselves, but they definitely lack the right to threaten members of society B. And society B has the right to defend itself against aggression from society A.

The Harm Principle as the Prime Directive

In applying the harm principle, it makes no difference that the subject is the interplanetary setting of a science fiction film. The principle is valid at any level. Take the Spanish civil war (1936–39). We might say that the residents of Spain have the right to fight among themselves about how Spain should be run, and other countries ought not to intervene. But let's say the Spaniards start fighting with weapons that threaten the Portuguese and the French. Then it becomes the business of the international community, or at least that of Portugal and France, to interfere, since Portugal and France have a right not to be threatened by Spain. This is precisely analogous to Klaatu's explanation to Professor Barnhardt. The other planets have been looking in on us and have noted our propensity to fight with each other, but now, we're developing both space flight and atomic weapons, either of which by itself would still not constitute a threat to other worlds, but which in combination present a real threat to innocent others. This, as Klaatu puts it, cannot be tolerated.

What does "cannot be tolerated" actually mean? Klaatu means it literally:

> Professor Barnhardt: Suppose this group should reject your
> proposals. What is the alternative?
> Klaatu: I'm afraid there is no alternative. In such a case, the planet
> Earth would have to be eliminated.
> Professor Barnhardt: Such power exists?
> Klaatu: I assure you, such power exists.

In other words, Klaatu's society is prepared to destroy Earth rather than let it threaten the peace and security of other worlds. This might be understood as a version of containment. In the Cold War, once it became clear that the Soviet Union would not be giving up communism any time soon, American policy makers had to settle for making sure that the Soviets did not export communism. The analogy isn't complete, however, for the Soviets and the Americans had roughly equivalent military capacities, whereas Klaatu's society has unilateral power to destroy a noncompliant Earth. So in this case, it would be a matter of Klaatu's society containing Earth's internecine problems so as to ensure that they do not threaten the security of others.

Such Power Exists

More precisely, it turns out that it is Gort's "society" that has this power. When Klaatu addresses the international group at the film's climax, he gives the full explanation of his mission:

> The universe grows smaller every day, and the threat of aggression by any group, anywhere, can no longer be tolerated. There must be security for all, or no one is secure. Now, this does not mean giving up any freedom, except the freedom to act irresponsibly. Your ancestors knew this when they made laws to govern themselves, and hired policemen to enforce them. We, of the other planets, have long accepted this principle. We have an organization for the mutual protection of all planets, and for the complete elimination of aggression. The test of any such higher authority is, of course, the police force that supports it. For our policemen, we created a race of robots. Their function is to patrol the planets in spaceships like this one and preserve the peace. In matters of aggression, we have given them absolute power over us. This power cannot be revoked. At the first sign of violence, they act automatically against the aggressor. The penalty for provoking their action is too terrible to risk. The result is, we live in peace, without arms or armies, secure in the knowledge that we are free from aggression and war. Free to pursue more profitable enterprises. Now, we do not pretend to have achieved perfection, but we do have a system, and it works. I came here to give you these facts. It is no concern of ours how you run your own planet, but if you threaten to extend your violence, this Earth of yours will be reduced to a burned-out cinder. Your choice

is simple. Join us and live in peace, or pursue your present course and face obliteration.

At first glance, their plan seems reminiscent of the argument by the seventeenth-century philosopher Thomas Hobbes that in order to keep the peace, the sovereign must have absolute authority over his subjects.[5] According to Hobbes, in our natural condition we are more prone to competition than to cooperation, and we come either to see predation as advantageous or to fear that others will think this. Although it might seem to everyone's advantage to agree to cooperate, we are not able to trust each other. Therefore, Hobbes concludes, we must invest all power in the sovereign, so that we are more afraid of breaking our social contract than we are of our neighbors. Our fear of the sovereign's power makes us law-abiding.

The key difference is that for Hobbes, the sovereign's power is far more extensive. As long as the sovereign is successful at keeping the peace, he cannot be called unjust, even if he inhibits travel, suppresses unpopular religions, or censors the press. Gort and his fellow robots, we're told, have jurisdiction only in matters of interplanetary aggression. So planet A might be a utopian community of benevolent poets and musicians, and planet B might be one of gladiatorial games and strife between factions, and in neither case could the robots interfere with how those planets' societies operate. The robots' Hobbesian authority arises only when one planet aggresses against another. Hence, although Klaatu's ultimatum is related to Hobbes's argument, its scope is substantially different, and one ramification of that difference is that the robots really are benevolent in a way that the Hobbesian sovereign need not be.

The robots are not paternalistic either, at least not in the traditional sense of regulating all our actions for our own good. It's true that we, and everyone else in the galaxy, will be better off if we eschew interplanetary nuclear war, so by prohibiting us from engaging in *that* pursuit, the robots' programming can be seen as paternalistic, but otherwise we are left free to regulate our own affairs, develop whatever social institutions we think best, and adopt any number of ethical codes. The security provided by the robots allows us the freedom to grow and develop, possibly into a mature and prosperous culture such as that from which Klaatu has been sent. The choice, as Klaatu says upon departing, is ours. To the greatest extent possible, then, our autonomy *is* being respected, and the greatest extent possible is that under which we can live without threatening our neighbors.

In one way, then, Klaatu's visit would not necessarily change our world—it wouldn't eliminate all tensions between the United States and the Soviet Union, or between Hutus and Tutsis. It would only force us to act responsibly toward other planets. In another sense, of course, it might change our world—by making us see the costs of war and conflict. After all, if the rest of the galaxy prospers by eschewing interplanetary war, perhaps Earth would prosper if we eschewed intercontinental war. As Klaatu notes, his people's freedom from aggression gives them freedom to pursue more profitable enterprises. That principle applies on the micro level as well as the macro level. He does mention, while talking to White House aide Mr. Harley (Frank Conroy), that his people "have learned to live without" stupidity. Perhaps the ultimatum, while not requiring us to rethink our priorities, would encourage us to do so.

Do the members of Klaatu's society have the right to deliver this ultimatum? I think that when we consider the actual context, they surely do. Again, they are not paternalistically telling us how to run our affairs. They are telling us that if we show signs of aggression that threaten their peace and security, they will retaliate. This is best characterized as preemptive self-defense. It is not the same thing as the preemptive strike in current military jargon, since the robots are not actually doing any preemptive striking. It is the ultimatum from Klaatu that is preemptive: don't commit aggression, or Gort will retaliate. It's true that the other planets have the right to exist in peace and security, and the robots protect that right. Or, to put it another way, aggressor planets lack the right to commit aggression, and the robots enforce this prohibition. As Klaatu notes, the only freedom lost in this arrangement is the freedom to act in ways we don't have a right to in the first place. For there cannot be such a thing as a right to violate rights. No meaningful and coherent conception of rights could include such a thing. It would be logically contradictory for Smith to have a right to do something and for Jones also to have a right to do something that would preclude Smith's exercise of his right. Jones's exercise of his ostensible rights would mean Smith did not in fact have a right. This is incoherent. Any coherent understanding of rights thus has to have the boundary condition that one's rights not entail the violations of the rights of others.[6]

To illustrate this on a smaller scale, consider our law against murder. We are all more secure in our persons when there is a prohibition against murder. This does seemingly eliminate a freedom, viz. the "freedom" to murder people. But since we do not have a moral right to murder people in the first place, we're not losing any freedom that we'd be entitled to miss. This is one way to interpret the word *harm* in the nineteenth-century philosopher John Stuart Mill's now famous argument that "the only purpose for which power can be rightfully exercised over any member of a civilized community, against his will, is to prevent harm to others."[7] If we read *harm* as "rights violations," then it's easier to see the inference that a system of rights cannot entail rights violations or, more simply, that there is no *right* to violate rights. But we can all equally and simultaneously enjoy a right to peace and security. Just as laws prohibiting murder do not violate anyone's rights, then, whether consented to or not, Klaatu's ultimatum does not violate anyone's rights. The moral foundation of Klaatu's ultimatum is that even if we do have the right to harm ourselves with our localized fighting, we do not have the right to harm those on other worlds, as they *do* have the right to peace and security.[8] Since we lack the right to do what Gort is forbidding us to do (to be aggressive toward other worlds), our rights are not violated by the ultimatum. The objection that this is coercive interference in our affairs presupposes that all our actions are presumptively rightful, but this is false, as we do not have a right to aggression. A toppled tyrant has no grounds for objecting to his ouster, since he had no right to his authority in the first place; similarly, were we to be prohibited from threatening other worlds, we could hardly object on the grounds that we ought not to be interfered with.[9]

Klaatu Barada Nikto

One possible interpretation of the allegorical significance of the robots might be to see them as our own nuclear weapons, which are often characterized as a deterrent. The idea is that we don't have to worry about attacks from other countries, since they know we have nuclear weapons. That seems to parallel Klaatu's point that his society is free from aggression because they know about the robots. One objection to this interpretation, however, is that our nuclear weapons are controlled by humans—fallible and passionate creatures who might not restrict their use in the stated manner—whereas Gort and his kind are neither fallible nor passionate. They respond only to the threat of interplanetary violence, and can be trusted to restrict the use of their power. The now-famous "safe word" expression "Klaatu barada nikto" that Helen Benson (Patricia Neal) uses to restrain Gort is likely a fail-safe feature installed for the purpose of a diplomatic mission such as Klaatu's. In the event of a "misunderstanding," Klaatu would need to have the ability to tell Gort to stand down.[10] But once the robots are patrolling, there is no safe word. Klaatu emphasizes that the robots' power in matters of policing interplanetary aggression is irrevocable. There is no "barada nikto" option, but presumably none would be needed. According to Klaatu, the robots work exactly as designed, using deadly force only when provoked by interplanetary aggression. Of course, it has since become a staple of science fiction that the machines charged with protecting us from ourselves will misuse or abuse their power. For instance, in *Colossus: The Forbin Project* (Joseph Sargent, 1970), a supercomputer is given control over the United States' nuclear missiles. It turns out that the Soviet Union has done the same, and the two computers merge, producing a regime of global security, but one in which individual freedoms are more strictly curtailed. The computer, Colossus, explains: "Under my absolute authority, problems insoluble to you will be solved: famine, overpopulation, disease. The human millennium will be a fact as I extend myself into more machines devoted to the wider fields of truth and knowledge. . . . We can coexist, but only on my terms. You will say you lose your freedom. Freedom is an illusion. All you lose is the emotion of pride. To be dominated by me is not as bad for humankind as to be dominated by others of your species." That's roughly similar to Klaatu's speech, but far more paternalistic. Klaatu suggests that we might learn to prosper and develop if we were less bellicose, but the bottom line is simply that we must not try interplanetary aggression. Colossus will insert itself into all fields of human endeavor, "solving all the mysteries of the universe for the betterment of man."

This theme reappears in an even more malicious form in *The Terminator* (James Cameron, 1984). Here, the computers that are in charge of coordinating nuclear defense decide that all humans are equally bad, not just the Communists, and should thus be exterminated. Kyle Reese (Michael Biehn), visiting present-day Los Angeles from forty or so years into the future, explains that "there was a nuclear war. A few years from now, all this, this whole place, everything, it's gone. Just gone . . . It was the machines. Defense network computers. New . . . powerful . . . hooked into everything, trusted to run it all. They say it got smart, a new order of intelligence. Then it saw all people as a threat, not just the ones on the other side. Decided our fate in a microsecond: extermination." The shift in attitude that is reflected by this contrast is interesting: in the 1951 case, the idea was that the robots could be trusted to ensure our safety, and as early as 1970, the primary object of fear was not the enemy but the technology run amok.[11] In any case, whether or not the machines are fallible, it's perfectly clear that people are fallible, which is why we worry about nuclear wars in the first place: in Sidney Lumet's 1964 *Fail-Safe,* it is human error that is primarily responsible for the destruction.[12] Regardless of whether we feel comfortable trusting Gort, part of Klaatu's point is that *we* cannot be trusted to play safely with nuclear weapons in an interplanetary context, because no one can be. The robots make sure no one is irresponsible with nuclear weapons technology.

It's certainly worth noting, of course, that the fact of our possession of nuclear weapons does imply the obligation to use them responsibly, just as any weapon needs to be used responsibly. It would be wrong, for instance, to use passersby as target practice for one's rifle. Nuclear weapons pose a far greater risk: massive devastation, both in terms of numbers of people killed and in terms of destruction of physical place. The more dangerous the weapon, the greater the need for responsible behavior. This is precisely what Klaatu is highlighting when he tells the scientists that the only way to achieve mutual security is to give up the "freedom" to act irresponsibly. We clearly do have an obligation to use our dangerous technology in wise and responsible ways, which includes eschewing aggression. *The Day the Earth Stood Still* is thus science fiction at its best, forcing us to take a look at ourselves and what we're doing by showing how it might appear to an alien. To Klaatu, it doesn't matter so much who rules Spain. But if we're going to start sending nuclear weapons into space, we're putting the peace and security of others at risk, and that's not something we are entitled to do. Klaatu's use of technology to achieve an ethical end highlights some of the dimensions of how we use dangerous technology.

Notes

I am grateful to Steven M. Sanders for offering many helpful comments on earlier drafts of this essay and for giving me the opportunity to write about this important film.

1. I am grateful to Steven M. Sanders for bringing this issue to my attention.

2. This and all other quoted dialogue, unless otherwise noted, is from *The Day the Earth Stood Still* (Robert Wise, 1951).

3. Interestingly, in Alan Moore and David Lloyd's seminal graphic novel *Watchmen* (New York: DC Comics/Warner Books, 1986), the scenario from *The Day the Earth Stood Still* is the inspiration for Ozymandias's plan, which *is* to stop us from fighting among ourselves. I discuss this in "Superhero Revisionism in *Watchmen* and *The Dark Knight Returns*," in *Superheroes and Philosophy,* ed. Tom Morris and Matt Morris (Chicago: Open Court, 2005).

4. A classic source for this idea is John Stuart Mill's *On Liberty* (1859; repr., New York: Penguin, 1985), which I discuss further below.

5. See Thomas Hobbes, *Leviathan* (1651; repr., New York: Penguin, 1984), chaps. 13, 14.

6. This is in the context of moral rights. Of course, contractual rights can be created in which only one person can do a thing at a time, for instance in the use of a timeshare.

7. Mill, *On Liberty,* 68.

8. For more on the notion of nonconflicting rights, see Hillel Steiner, "The Structure of a Set of Composible Rights," *Journal of Philosophy* 74 (1977): 767–75.

9. For further discussion of this idea, see my "War and Liberty," in *Reason Papers* 28 (2006): 43–49.

10. Klaatu's people, or the robots, must have a remarkably efficient language if three short words can convey the instructions "Abort your normal procedure of laying waste to this planet and instead come find me and bring me back to the spaceship so you can revive me from my injuries. In case this message is delivered by a surrogate, bring her also."

11. Of course the theme of technology run amok has always been a staple of science fiction, dating to at least Mary Shelley's 1818 *Frankenstein.* But post-Sputnik American science fiction typically shows an optimism about technology that, by the 1970s, seems to have receded.

12. *Fail-Safe* is, I would argue, an example of science fiction, despite the absence of anything "futuristic."

Some Paradoxes of Time Travel in *The Terminator* and *12 Monkeys*

William J. Devlin

I'll Be Back (to the Future) . . .

Suppose you had a time machine. Where exactly would you like to go throughout all the possibilities of temporal locations? Would you want to go back to the Jurassic period to learn more about the dinosaurs? Maybe you would like to go back to ancient Greece to finally know whether or not the Battle of Troy really took place. Perhaps the past bores you, and you're really a future adventurer instead. If so, would you fast-forward to 3050 to see if human beings are riding in flying cars and living on the moon? Maybe you'd like to go even further, say to 8000, to see how far the human species has developed (or if we've even survived).

When we ask ourselves such hypothetical questions concerning time travel, we usually don't think of temporal locations to visit that are so vastly distant that they would have very little effect on us. On the contrary, quite often our ideal time voyages include a temporal place that we would like to change. This change can be something that has an immediate effect upon our past before we get back to the present, or a mediated effect that can take place for our future once we return to the present. For instance, you may want to change the past by going back to 1889 to kill Hitler's mother before he was even born, thereby possibly eliminating World War II and changing the course of history in the twentieth century. Or you may want to benefit your own future by going back to 1986 to invest in Microsoft so that when you return to the present, you can be a millionaire, financially set for the rest of your life.[1]

We find this incentive to use time travel to make significant changes, whether they are changes that take place in the past or in the future, in two popular science fiction films. The first film, James Cameron's *The Terminator* (1984), gives us a case of changing the past in such a way that it has an immediate effect upon the given timeline. The film tells us of time traveling from the postapocalyptic future that centers around a war between humans and machines. In 1997, Skynet, a computer system that gains its own intelligent independence, wages war on human beings and wipes out nearly all of humanity. A small human resistance, led by John Connor, is able to rise up and, over three decades of war, come to defeat Skynet and the machines. Realizing its imminent defeat in 2029, Skynet sends the cyborg Terminator back in time to 1984 on a mission to kill Sarah Connor, the mother of the yet to be born John Connor, so as to eliminate the leader of the resistance. Here, Skynet maintains that if John Connor is eliminated, the course of history from 1984 to 2029 will drastically change, allowing Skynet and the machines a better chance to both defeat human resistance and dominate the world. John, aware of Skynet's plan, sends his friend and fellow officer, Kyle Reese, back to 1984 to protect Sarah from the Terminator. The war of the future thus continues in 1984.

The second film, Terry Gilliam's *12 Monkeys* (1995), provides us with an example of using time travel to the past to make changes that can take place only in the time traveler's future. This alternative adventure from the postapocalyptic future centers around the outbreak of a lethal virus that wipes out 5 billion people in 1996, leaving only 1 percent of the original population to survive underground by 2035. In this dark and miserable underground society, a panel of leading scientists searching for a cure sends volunteers back in time prior to the outbreak to gather information. James Cole, a convict and sociopath, is selected by the panel to go back to 1996 both to learn about the Army of the 12 Monkeys, which is suspected to be responsible for spreading the epidemic, and to help secure a sample of the virus before it mutates so that the panel can study it and find a cure. While in 1996, Cole must struggle through his (1989) psychiatrist's view that he is insane so that he can complete his mission, which will not alter the already given future events between 1996 and 2035 but will provide information to the scientists so that they can work toward bettering humans' uncharted future.

Both films treat their viewers to a voyage through one of the most fascinating and intriguing ideas in science fiction literature: namely, the idea of time travel. Addressed in books, films, and TV shows, time travel is a popular theme in the science fiction corpus, a theme that is explored from many different angles: from the ways in which one can travel back through time to the wide variety of things one can do once one travels in time. But while many types of questions arise in the fantasy realm of this literature, several important philosophical questions concerning time travel will be addressed in this essay. These questions center on personal identity, freedom, and logical paradoxes for both the time traveler and the world he lives in. Is there an identity crisis if I go back in time and meet myself? Does time constrain my freedom of action? Can I go back in time and kill my grandfather before he meets my grandmother? If so, what happens to me? Can I really change events in the past? How is this possible if the events have already occurred? In this essay I explore how the two films address these philosophical questions regarding time travel. The possible answers to these questions reflect two different approaches to making changes when one travels back in time. Let's see how these films work through the questions.[2]

Building Our Own Time Machine: A Beginner's Manual

When we think about traveling through time, there are several problems, or even paradoxes, that the time traveler may face, whether they are paradoxes that confront him personally, paradoxes that hold for the whole universe, or both. We can break down these problems into two categories of paradoxes. The first type is the *empirical paradox,* which concerns the scope of paradoxes that stem from the experiences of the perspective of the time traveler. Here, we are interested in questions about what may happen to one's sense of self in time travel, the effects changes in the past may have upon oneself, etc. The second type is the *metaphysical paradox,* which concerns the scope of paradoxes that stem from logical impossibilities that arise from traveling through time. In this category, we deal with such issues as the ability to eliminate one's own past self, the problem of finding an original cause in circular causal chains, etc.

Our exploration of the philosophical issues regarding time travel in *The Terminator* and *12 Monkeys* will follow these two categories. However, before we begin examining the details of each of the questions regarding time travel, we need to first flesh out what we mean by time travel. It is one thing to travel from one spatial location to another, such as when you fly from Boston to London or drive from Wyoming to California. But what exactly do we mean by traveling from one temporal location to another? We can follow philosopher David Lewis's definition of time travel, which holds that someone travels through time if and only if the difference between the traveler's departure and arrival times in the surrounding world does not equal the duration of the traveler's journey. So, in *12 Monkeys,* Cole can be understood as a time traveler if he travels through the time machine in 2035 for two hours, which brings him to 1996. Here, the difference between Cole's arrival and departure times is thirty-nine years, while the duration of his journey is only two hours, and so it is clear that Cole has, indeed, traveled through time. The time traveler is thus one who moves from one temporal location (departure) to another temporal location (arrival), whether it is a past or future location, and the difference between these times is distinguishable from the time the traveler takes to reach his destination.

While this definition helps us to see what we mean by *traveling* through time, the question of time still remains. After all, it seems as though Lewis's account of time travel includes two distinct accounts of time: *external time,* or the objective account of time in the natural world, and *personal time,* or the subjective account of time in the traveler's personal journey. What exactly, then, is time? How do we go about explaining what time is?[3]

One method is to appeal to physics to ground the notion of Lewis's external time. Here, we find that several different accounts of the nature of time arise in modern physics. The Newtonian account of the universe holds that time, like space and motion, is absolute: one can measure the interval of time between two events, and this measurement is the same for anyone else who measures such an interval. According to the Einsteinian account, special relativity maintains that since the laws of physics are the same for any inertial observer (or observer whose trajectory has a constant velocity), the space-time interval between two events is common for all inertial observers. But time is not absolute, since the same event can happen at different times for the stationary inertial observer and the inertial observer moving toward a different observer. The measurement of space-time interval is thus dependent upon the observer's frame of reference. Meanwhile, general relativity expands on special relativity, as it includes noninertial frames (or frames of reference that do not have a constant velocity) and introduces the notion that space-time is not linear but curved.[4]

Another approach to understanding what we mean by time is to turn to philosophical accounts of time. Here we find three distinct views about time. The first view, *eternalism,* maintains that time is a fourth dimension that, together with the three dimensions of space, constitute the essential aspects of reality. The three elements of time—past, present, and future—are treated as real, as are all space points of the space-time diagram. This entails that all past and future temporal locations, objects, and events have the same ontological status as eternally present objects and events. For instance, ontologically, Jeffrey Goines of 1996 and his leadership in the Army of the 12 Monkeys is as real as Cole of 2035 and his time spent in prison, even though from Cole's 2035 perspective, Jeffrey is dead. Likewise, Sarah Connor of 1984 and her life as a waitress are just as real as the future John Connor and his leading role in the resistance against Skynet, even though from the perspective of Sarah Connor, her son hasn't even been born. This equal ontological eternal reality for all temporal locations entails that all locations are fixed and cannot be changed. The second view, *possibilism*, accepts the eternalist's view concerning the ontological status of the past and present but rejects the ontological status of future events. That is, the future is not fixed but is open to change, so that there are a variety of possibilities. Whether it is the survivors of the virus outbreak who are trapped underground in 2035 or even Skynet and the machines that are losing the battle against the human resistance in 2029, the future possibilities are open; no future event has yet been actualized. The third view, *presentism*, challenges both the eternalist and possibilist positions by maintaining that only present objects and events are real. Although the past objects and events did once occur, they no longer exist. As the future objects and events have yet to be actualized, they too do not exist. For Cole and those who presently live in 2035, the outbreak of the lethal virus of 1996 is not real. For Sarah Connor and those who presently live in 1984, the eruption of war between Skynet and human beings in 1997 is not real.[5]

Thus both the scientific and philosophical approaches to defining time give us a variety of answers to the question, What is time? While the scientific approach to defining time will help in discussing the physical possibilities of time travel, the philosophical approach will help in exploring the logical possibilities and empirical and metaphysical issues that arise with our time travelers, Cole, Reese, and the Terminator. So let's keep the three distinct philosophical views—eternalism, possibilism, and presentism—in mind as we turn to some philosophical issues of time travel.

How Many James Coles Can There Be?

The first philosophical issue falls under the category of empirical paradox; it concerns the question of personal identity in time travel. Particularly, the question centers on the empirical problem of encountering yourself when you travel back in time. We see this problem in *12 Monkeys* when Cole of 2035 returns to 1996 for his last time and joins his psychiatrist, Dr. Kathyrn Railly, at Philadelphia International Airport. We know that the young Cole of 1996 was present at the same airport on the same day in 1996 and even witnessed the older Cole's death. Thus it appears as though we have two Coles, the younger Cole (call him $Cole_1$) and the older Cole (call him $Cole_2$).

This claim, however, runs into a contradiction, as it violates the *principle of indiscernibility of identicals*, or what is known as Leibniz's law, which holds that if what appear to be two or more objects are identical, there can be no property held by one and not the other(s). That is, $Cole_1$ and $Cole_2$ may appear to be two different people, but they are in fact the same person, James Cole. Yet $Cole_1$ and $Cole_2$ have different properties: $Cole_1$ is younger, shorter, and has hair, while $Cole_2$ is older, taller, and bald. This leads to a paradox insofar as the single individual, James Cole, has the contradictory properties of being young and old, having two different heights, and having and not having hair. Thus, if we have one person, we violate Leibniz's law. But if we have two different persons, then we cannot consistently say that both people are, in fact, James Cole.[6]

We can resolve this problem by following Lewis, who focuses on the individual time traveler's personal time in relation to his personal identity to overcome the empirical paradox of meeting one's self in the past. We can say that one's personal identity consists of a continuity of both mental and physical states, where each set contains states that are continuous via causal relations. For Lewis, we can focus on the mental states in marking the changes of one's identity through time travel. One's mental states, or cognitions, undergo changes exhibiting causal continuity. But mental states occur from one's personal perspective. Now while normally, for non-time travelers, our personal perspective of mental states and changes in cognition follows both our subjective personal time and objective external time (since they coincide), the time traveler will notice a difference in times concerning identity. Namely, once the two times separate, we realize that our personal perspective of mental states and changes of cognitions follows a causal continuity along our subjective personal time. The time traveler may jump to different and disparate external times, but his mental states follow the continuous temporal line of personal time. For Lewis, then, the identity of the time traveler is consistently and causally continued along the traveler's personal time.

Lewis's emphasis on the relation of personal identity to personal time resolves the paradox of personal identity for Cole. The event of $Cole_1$ and $Cole_2$'s being at the same airport at the same time is an event for just one person—James Cole. From the objective perspective in external time, we can say that James Cole has two different bodies at this event, $Cole_1$ and $Cole_2$. But from the subjective perspective of personal time, Cole is one person who has his own distinct, causally continuous cognitive experiences of perceptions, which include perceptions of the other Cole. $Cole_1$ is the young James Cole who has the personal experience of perceiving $Cole_2$'s being shot and killed. $Cole_2$ is the older James Cole who runs by $Cole_1$ and who may recall being in $Cole_1$'s shoes earlier in his life. While it is true that these two Coles have different mental states, they can still be the same person and not violate Leibniz's law since these different mental states are not duplicates of mental states of the same stage according to the external time of 1996. Rather, since we are following Cole's personal time, which is causally continuous and can be consistently traced from $Cole_1$ to $Cole_2$, $Cole_1$ and $Cole_2$ can both be the same Cole at different stages of Cole's life. Thus while from the objective perspective of external time it may appear that there are two different persons, when we focus on personal identity from the subjective perspective of personal time, there is only one Cole, who happens to exist at a certain external point in time twice.

Can I Terminate My Past Self?

With the empirical paradox of personal identity resolved, we can see that the time traveler has a great deal of freedom in traveling through time. He has the liberty not only to go back to a time frame where his younger self exists but also to join his former self in the same event, and even talk to him if he wishes. But how far does the time traveler's freedom go? The question of what a time traveler can and cannot do brings us to one of the most challenging logical paradoxes, which we can imagine occurring in both of our science fiction films. In scenario one, suppose that, in 2029, instead of leading the human resistance to victory over Skynet and the machines, John Connor gradually goes mad in this postapocalyptic tumultuous world. In his madness, when he learns that Skynet has sent the Terminator to 1984 to kill his mother, John decides to outdo the machines by going back to 1983 himself to kill his own mother before the Terminator can. In scenario two, imagine that, in 1996, as Cole chases Dr. Peters with the intent to kill him with his revolver, Cole slips on a banana peel. As he falls to the ground, his gun goes off and kills the younger Cole who is an onlooker to the chase.

Both scenarios give us the case of the logical paradox called the *grand-father paradox*. This problem centers on the issue of whether or not a time traveler can go back in time and perform certain actions that eliminate his past self and his existence. If John Connor murders his mother, Sarah, in 1983, he will stop her from conceiving any children. This means that John will not be born. But if John will not be born, how can he exist in 2029 and go back in time to 1983? Likewise, if Cole of 2035 accidentally kills his younger self of 1996, Cole ceases to live in 1996. Given what we've said about personal identity in one's personal time, the lifeline of Cole ceases in Cole's personal time of his first presence in 1996 as a young boy. But if Cole dies as a young boy in 1996, how is he able to exist in 2035 and return to 1996?[7]

How are we to evaluate this predicament? Can John murder his mother before he is born? Can Cole kill his younger self before he grows up? Both actions appear to be impossible, given the causal continuity between the time traveler and the person who is killed. But, at the same time, the actions of murder and accidental killing are physically possible. Both John and Cole are quite capable of killing people. We thus have a dilemma. On the one hand, if we maintain that it is logically impossible for John and Cole to directly or indirectly terminate their younger selves, then we must account for why they physically cannot do it. Why can't John take a gun and pull the trigger while aiming at his mother? Why can't Cole's gun go off and kill his younger self? On the other hand, if we allow for either one to perform such actions, we need to account for the fact that they still exist.

When we turn to each film for a response to this paradox, we see that they give us different answers, which depend upon their ontological accounts of time and time traveling. Let us take the latter case of Cole's slip on the banana peel first. As suggested in the opening section, the time traveling premise of *12 Monkeys* is that the time traveler is one who can travel back to the past to help make changes only for the future. The time traveler cannot change the past in such a way that it creates a new present. The past and present are fixed since the events have already happened or are happening at this very moment. The only temporal location that can be changed is the open future, since those events have not yet occurred. Cole himself points this out to the panel of psychiatrists in 1990 when asked if he is there to save them from the virus outbreak of 1996: "Save you? How can I save you? It already happened! I can't save you. I'm simply trying to get some information for people in the present [2035] so that someday . . ." Following Cole's view of time travel, we can see that the film adopts the possibilist view of time—the past and present are fixed and unchangeable, while the future is still open. More specifically, the film presupposes the *growing universe* metaphysical model, where there exists a single four-dimensional trunk that contains the temporal locations of the past, while the present is located at the edge of the trunk. The future, however, as it does not yet exist, is not in or on the trunk and so can be brought forth in many different ways.[8]

With this view of time in mind, can Cole slip on a banana peel and accidentally kill his younger self? Well, the best answer for the picture of metaphysics in *12 Monkeys* is that Cole can and cannot kill his younger self. Here, we will use Lewis once again to help us solve the problem. As suggested in the paradox, it is possible for Cole to accidentally kill his younger self—we can all imagine slipping while running and dropping something that we are carrying. The scenario is not a physical impossibility. Likewise, it is even possible to imagine Cole intentionally turning his gun on his younger self—he may be so confused by his coexistence in 1996 and 2035 that he decides to end the charade and murder his younger self to test his sanity. This alternative scenario is also physically possible. Thus, in one sense, Cole *can* kill his younger self—he has the ability, the intent, and the opportunity. However, in another, more important sense, Cole *cannot* kill his younger self. Given the possibilist–growing universe model of time, it is a fact in the past that young Cole was at the airport and was not killed, either accidentally or intentionally. And, since the past cannot be changed, the older Cole cannot create new events in the past. This sense of *cannot*—the sense that the past is fixed and unalterable—concerns the logical impossibility of Cole's changing the past by killing his younger self. This latter, logical sense of *can* and *cannot* explains why Cole ultimately cannot perform such an action, even though he has the physical ability to pull the trigger and aim his gun. While the traveler is free in the sense that he has the ability to do many things, his actions are shaped to follow the directions of the past events as they unfolded. If Cole goes back to 1996, while his actions in the past are new to him from his subjective perspective, they are actions that have already occurred from the objective perspective of external time. And since the past is fixed, they cannot be changed.[9]

While *12 Monkeys* disarms the grandfather paradox by ultimately saying that Cole cannot kill his younger self because he cannot change the past, *The Terminator* takes a different approach to resolving the paradox. When we turn to philosophical views of time, we find that *The Terminator* holds the same view of time as *12 Monkeys:* the possibilist view. Like *12 Monkeys, The Terminator* operates on the assumption that the future is open to change. But *The Terminator* alters and expands this view when it comes to the kind of model of the universe it adopts. *The Terminator* espouses the *branching universe* metaphysical model, where, as in the growing universe model, there is a single four-dimensional trunk that contains the temporal locations of the past, while the present is located at the edge of the trunk. Meanwhile, the future is open and is marked by an unactualized set of branches that stretch out beyond the present. Each nonactual branch represents one possible way that things can unfold. But while the growing universe model maintains that the past and present are fixed and cannot be changed, the branching universe model can be construed so as to allow for the possibility that the past and present can be changed because there are possible branches along the past and present that link to alternative world timelines. The time traveler can leave his present time, travel back into the past, and make changes that can not only affect the future of the time traveler's world but also change the course of events that follow immediately from the changes made in the past. Such changes move the time traveler, and the world around him, to a branch that leads him to an alternative world.[10]

We can see this possibilist–branching universe model of time at work in *The Terminator* when Skynet sends the Terminator back to 1984. Whereas Cole is sent back to 1996 to retrieve information to help the future beyond 2035, Skynet wants to do more than gather information to help its future beyond 2029. Aware of its imminent defeat in the war against human beings, Skynet wants to save itself and the machines by changing history. Killing Sarah Connor will eliminate the leadership of John Connor. This elimination will change the course of events that unfold when Skynet attempts to control the world. Skynet will now have a better chance of winning the battle against the humans so that its loss of the war in 2029 won't even occur. Instead, Skynet and the machines will hold power and control over the world in an alternative temporally present location. *The Terminator* thus assumes that the past can be changed so that changes in the past will lead to a new history and new present.[11]

Using the possibilist–branching universe model of time, how would *The Terminator* resolve the grandfather paradox? Can a mad John Connor go back to 1983 and kill his mother? Like *12 Monkeys, The Terminator* would follow Lewis's first point, that John can kill his mother insofar as he has the ability, the intent, and the opportunity. However, *The Terminator* would challenge Lewis's second point, that John cannot kill his mother since the fact that he did not already do it in 1983 (as Sarah gives birth to John after 1983) cannot be changed. Given the branching universe model, the past can be changed. Thus John can kill his mother in both senses of the word *can*—John has the physical ability to do it and it is not logically impossible. Furthermore, this model helps to disarm the paradox on its logical side by allowing for the movement into an alternative world timeline. John can be born in the world in 1984, enter the time machine in 2029 to go back to 1983, kill his mother, and thereby change the past so that he creates an alternative 1983 where his mother is dead and he does not come into existence. Given the existence of, and bridge between, two worlds, John can follow through with such an action that can be consistently connected throughout his existence and not encounter the logical contradiction found in the grandfather paradox.

Caught in the Circle of the Army of the 12 Monkeys

Thus far, we have seen empirical and logical paradoxes arise in time travel through *The Terminator* and *12 Monkeys* that can be resolved in one way or another. But there is one prominent logical paradox that remains, and it can be found in *12 Monkeys*. In 2035, the scientists searching for a cure for the lethal virus obtain a voice mail from 1996 informing them that the Army of the 12 Monkeys is responsible for the virus outbreak and so can help them find a cure. The panel plays this message for Cole so that he knows that he must find the army in 1996. When Cole returns to 1996, he informs Dr. Railly that he is searching for the army and tells her the telephone number that he is to call to check for voice mails that may have further leads. After deciding that she and Cole are insane, Dr. Railly calls the number and jokingly leaves a message stating that the Army of the 12 Monkeys is responsible for the outbreak of the virus. To both Dr. Railly and Cole's bemusement, it turns out that that message is the message the panel of scientists received in 2035.

Now, the question is, where did this information—that the Army of the 12 Monkeys is responsible for the outbreak of the virus—originate? Following the timeline of the movie, the information is passed along through four events: (1) the panel of scientists receives it from a voice mail, (2) Cole receives the information from the panel of scientists, (3) Dr. Railly receives the information from Cole, and (4) Dr. Railly reveals the information on a voice mail that is later picked up by the panel of scientists. The problem is that while the four events form a causal sequence so that each event can be causally explained by the prior event, the information cannot be causally explained to have a beginning or an end. This paradox is known as a *causal loop*, where the information causally flows in a circle without a definite point of origination.

There are two different ways to respond to this problem. First, we can admit that causal loops are impossible, whether they are loops in a normal, external timeline or in time travel timelines. Here, the argument maintains that it is impossible to have an uncaused cause, and since the information passed along from the panel of scientists to Cole to Dr. Railly to the voice mail has no origination, this information is a case of an uncaused cause. And since this information is the essence of the causal loop, the causal loop is impossible. If this is the case, then *12 Monkeys* must give an explanation for the origination of the information by providing an independent source that is outside the causal sequence. For instance, the panel of scientists may have retrieved a newspaper from 1996 with the headline "The Army of the 12 Monkeys Did It!" and (mistakenly) inferred that *it* refers to the virus outbreak. The panel then uses this source as its primary source for the information and uses the voice mail as corroborating evidence. In this case, the information can be passed along to Cole (and to Dr. Railly and to the voice mail) without falling into a causal loop (providing that none of the events in the causal sequence causes the headline). The second response is to maintain that uncaused causes are indeed possible. For instance, one may wish to hold that the universe, or the big bang, has no original cause. If we accept this point, then the information passed along in the causal sequence can be allowed as an uncaused cause. Therefore, while *12 Monkeys* does present a series of events that follow a causal loop, this loop is not a problem since uncaused causes are not logically impossible.

The Times They Are A-Changin'

As we've seen, both *The Terminator* and *12 Monkeys* take their viewers through time travel adventures into the past. Whether it's through the Terminator, whose mission is to change the course of history, or James Cole, whose goal is to help change the future, both films espouse the idea that time travel can make changes in certain temporal locations. These two films, which center on time changes, use their suspenseful tales of traveling through time not only to raise hypothetical science fiction questions but also to broach serious philosophical questions concerning time travel. Whether it is the empirical paradox concerning one's personal identity when a time traveler meets his past self or the logical paradoxes of one's ability to kill one's past self and of allowing for uncaused causes, *The Terminator* and *12 Monkeys* provide a science fiction context for raising these philosophical problems and for proposing resolutions. Though they may disagree over what can be changed and what the metaphysical picture of time looks like, together the two films help us to further clarify the conceptual landscape of the philosophy of time travel.

Notes

1. We can find such a scenario of changing the historical events of the twentieth century by eliminating Hitler in Stephen Fry's comedic science fiction novel *Making History*, which tells us the story of Michael Young, a PhD student with a dissertation on Hitler's childhood who travels back in time to drop male sterility pills in the drinking water used by Hitler's parents. Michael then returns to his present time to deal with the changes that have been made now that Hitler has been eliminated.

2. While the entire corpus of science fiction time travel is too large to list, literary works include H. G. Wells's *The Time Machine;* David Gerrold's *The Man Who Folded Himself;* Lewis Carroll's *Through the Looking-Glass;* T. H. White's *The Once and Future King* and *The Sword in the Stone;* Robert A. Heinlein's short stories "By His Bootstraps" and "All You Zombies"; Ray Bradbury's short story "A Sound of Thunder"; Isaac Asimov's *The End of Eternity;* Jack Finney's *Time and Again, From Time to Time,* and *The Third Level;* Nicholas Meyer's *Time after Time;* James P. Hogan's *Thrice upon a Time;* Harry Turtledove's *The Guns of the South;* Kurt Vonnegut's *Timequake;* Michael Crichton's *Timeline;* and Terry Pratchett's *Thief of Time.* Time travel films include *All Over Again; Army of Darkness; Austin Powers: The Spy Who Shagged Me;* the *Back to the Future* trilogy; *Bill and Ted's Excellent Adventure* and *Bill and Ted's Bogus Journey; Carnivale; Clockstoppers; Donnie Darko; Dreamcatcher; Event Horizon; Le Jetée; Journey to the Center of Time; Minority Report; My Science Project; The Philadelphia Experiment; Somewhere in Time; Time Bandits; The Time Machine; The Time Travellers; Timecop;* and *Timeline.* Television shows that are centered on, or include stories of, time travel include *Buck Rogers in the 25th Century; Crime Traveller; Doctor Who; Futurama; The Outer Limits; Quantum Leap; Red Dwarf; Star Trek; The Time Tunnel;* and *The Twilight Zone.*

3. For further reading on Lewis's account of time travel, see David Lewis, "The Paradoxes of Time Travel," *American Philosophical Quarterly* 13 (1976): 145–52.

4. For further discussion on the scientific views of time in physics, see Stephen Hawking, *A Brief History of Time* (New York: Bantam Books, 1988).

5. It is not clear whether or not time travel is permissible from the presentist's view. For a discussion on this debate, see William Godfrey Smith, "Traveling in Time," *Analysis* 40 (1980): 72–73, and Simon Keller and Michael Nelson, "Presentists Should Believe in Time-Travel," *Australasian Journal of Philosophy* 79 (2001): 333–45.

6. For further discussion on the problem of personal identity in time travel, see Keller and Nelson, "Presentists Should Believe."

7. For a detailed account of the grandfather paradox, see Lewis, "Paradoxes of Time Travel"; Peter J. Riggs, "The Principle Paradox of Time Travel," *Ratio* 10 (April 1997): 48–64; and Theodore Sider, "A New Grandfather Paradox," *Philosophy and Phenomenological Research* 57 (March 1997): 139–44.

8. For further readings on the growing universe model and time, see Michael Tooley, *Time, Tense and Causation* (Oxford: Clarendon Press, 1997), and Kristie Miller, "Time Travel and the Open Future," *Disputatio* 1 (November 2005): 223–32.

9. We see several examples of this "rule" in the film, as Cole's actions into the past have effects upon people in the past, even though he had not done them yet in his personal time. For instance, from Cole's personal time, Dr. Railly has a photo of Cole from World War I in 1917 before he time traveled to that location; in 2035, Cole hears the telephone message about the Army of the 12 Monkeys before he is with Dr. Railly when she leaves that message in 1996; young Cole perceives his own death and grows up to follow the exact same steps of his death; etc.

10. For further readings on the branching universe model, see Hugh Everett, "Relative State Formulation of Quantum Mechanics," *Review of Modern Physics* 29 (1957): 454–62; David Albert and Barry Loewer, "Interpreting the Many Worlds Interpretation," *Synthese* 77 (1988): 195–213; Murray Gell-Mann and James B. Hartle, "Quantum Mechanics in the Light of Quantum Cosmology," in *Proceedings of the 3rd International Symposium on the Foundations of Quantum Mechanics in the Light of New Technology*, ed. S. Kobayashi, H. Ezawa, Y. Murayama, and S. Nomura (Tokyo: Physical Society of Japan, 1989), 321–43; Storrs McCall, *A Model of the Universe: Space-Time, Probability, and Decision* (Oxford: Clarendon Press, 1994); G. C. Goddu, "Time Travel and Changing the Past (or How to Kill Yourself and Live to Tell the Tale)," *Ratio* 16 (March 2003): 16–32; and Miller, "Time Travel."

11. To be sure, *The Terminator* trilogy moves back and forth between the possibilist–growing universe and possibilist–branching universe models of time. In *The Terminator*, although Skynet assumes the possibilist–branching universe model, the fact that Reese becomes the father of John Connor suggests the possibilist–growing universe model, insofar as it had to be the case that Reese is the father. Meanwhile, *Terminator 2: Judgment Day* (James Cameron, 1991) espouses the possibilist–branching universe, as Sarah Connor concludes that the future (which is still the past from the present year of 2029) is now open, since the pieces of the Terminator believed to be the catalyst for

the creation of Skynet have been destroyed and Dr. Miles Dyson, the creator of Skynet, is now dead. But in *Terminator 3: Rise of the Machines* (Jonathan Mostow, 2003), we once again return to the possibilist–growing universe model, as John Connor realizes that his failed attempt to stop Skynet has led him into a shelter that will allow him to regroup and eventually lead the human resistance in the exact same way as history has suggested.

2001: A PHILOSOPHICAL ODYSSEY

Kevin L. Stoehr

In *2001: A Space Odyssey* (1968) we are invited by director Stanley Kubrick to experience a mesmerizing yet also alienating form of sensory liberation, as paradoxical as such an experience may at first sound. His landmark science fiction film does not attempt to free us somehow from our five senses, certainly. In fact, the film tends to enhance an appreciation of our perceptual faculties, particularly those of vision and hearing, as well as to encourage reflection on what we have experienced through our senses while watching the film. But Kubrick's masterwork leads us beyond the borders of our conventional world of familiar perceptions and invites us to ponder abstract questions and ideas that seemingly transcend the boundaries of the sensory and perceptual world of everyday human experience. Most important, *2001* allows us to feel the type of disconnected, disembodied existence that results from any attempt—especially via technological means—to transcend the natural world and to replace the realm of concrete objects and events with a realm of indirect impressions, generalized moods, and sheer abstractions.

Now by virtue of the simple fact that it is a film, *2001* cannot avoid the human need for physical sensations and the world of particulars—in this case, the world of specific images and sounds that point beyond themselves to actual objects that cause such phenomena to emerge in relation to a perceiving subject. However, images and sounds also point beyond themselves in a more conceptual and intellectual way, one that certainly occurs when they are used or experienced metaphorically, symbolically, and philosophically, as is the case in Kubrick's film. We normally think of our familiar world of concrete physical objects and particular events as our natural world, but there are certainly kinds of environments—such as those in a cinema or in a space capsule—that are anything but natural, since their very function presupposes an attempt to provide sensations and perceptions that are machine made rather than nature given.

In a film whose central and most dramatic section deals with a machine-governed world, not to mention a film that owes much of its beauty and spectacle to cinematic technology, it should not be surprising that one of its major themes is the loss of the natural environment through technology, along with a corresponding reduction of the role played by our bodily senses and a subsequent sense of disconnection or dislocation. Like any film, *2001* provides an experience in which we, as film viewers, continue to rely upon our senses and physical bodies, though we do so in a way that makes us passive and separate from the natural world, something like an astronaut who perceives his environment only by staring through his space helmet or through a spaceship window. But Kubrick's film makes this sense of passivity and detachment into a central theme, in terms of both form (cinematic technique and style) and content (narrative).

2001 is a film about many things that are worthy of reflection and speculation: the dangers of technology, the mystery and sublimity of outer space, the fragility of humankind, the evolution of our species over time, the concept of intelligence, and so forth. Such themes have been explored and analyzed from various viewpoints since this motion picture was first released. It is a film that operates, according to Kubrick's own professed intentions, at the level of myth and metaphor and that thereby invites its audience's creative speculation. But above all else—since the possibilities of interpreting the film are seemingly endless, like the very expanses of space that it depicts—*2001* is a film that, simply put, takes its viewer to places that are no longer earthbound or tied directly to a familiar environment. This is especially the case given the contrast between the last two parts of the film and the earlier part of the film dealing with prehistoric life on Earth. And so conventional modes of location, direction, and connection are no longer in operation, causing the viewer to feel weirdly liberated and yet also deeply disconnected. It is as if the film viewer has somehow shed his body and yet nonetheless maintains the power of his five senses. In this theoretically possible but practically impossible manner, the viewer might still perceive detachedly all that soars past his fleeting and fluctuating center of consciousness, yet without being anchored by the confines of his natural and material embodiment. Kubrick's film engenders this hypothesized type of feeling.

A reflective viewer's experience of *2001* is a perfect illustration of the fact that images and sensations can give rise to ideas and thoughts that are no longer defined merely by our physical existence in space and time. The general concept of a space odyssey, for example, is much more than any particular physical journey through outer space, especially when it is presented in the allegorical and symbolic way that Kubrick has done here. Again, the film's very requirement of sensory perception on the part of the viewer might hint at the fact that our need for sensory perception can never be completely surpassed. And though the mind may occasionally play with images on its own, as when we dream while asleep, we require our physical body and its five senses in order to gain the very perceptions from which our thinking is born and from which it develops.

Simply put, despite our power to think conceptually and speculatively, we cannot deny the importance of our senses and their origins in the material existence of our bodies. Our knowledge is rooted in our perceptions, and our perceptions are rooted in the physical world of sensation, a world in which we always find ourselves as situated, located organisms. While Kubrick may play with the idea of some cosmic form of intelligence that is not rooted in any particular body, and while he even may suggest that humans might eventually become part of such an intelligence in a way that does not require physical embodiment, the film itself seems to invite reflection on the possibility that an attempt to transcend the body's anchorage in an earthly environment is, in fact, an attempt to overcome our very humanity. *2001* thereby raises the perennial philosophical question, What does it mean to be human?

Minds and Bodies, or Lack Thereof

We naturally toy at times with the idea of transcending our physicality through the power of our mind or spirit, and science fiction provides a perfect arena in which to imagine how our technology might assist in this endeavor, even if such an attempt must fail in reality. In *2001*, Kubrick presents a vision of the evolution of humankind to a point where mind takes precedence over matter and where the gods (i.e., extraterrestrial forms of intelligence) have used their own machinelike tools (in the shapes of black monoliths) to push humans past the confines of their own physical bodies and therefore past the restrictions of space and time. One reason for thinking that the advanced being or beings of *2001*, those that are responsible for the monoliths and their ability to activate intelligence in various parts of the cosmos, are disembodied or nonphysical beings is the fact that we never see or perceive such beings in the film. We merely witness their signs and tools, testaments to their actual though incorporeal existence.

2001 is not necessarily a celebration of the idea of any evolution or advancement toward such a form of existence, since the dangers inherent in this type of transcendence (i.e., the surpassing of the need for physical embodiment) are also evoked in the film. Most important, the viewer is taken on a journey that increasingly indicates the losses involved in forgetting the importance of our earthly bodies. While *2001* certainly evokes many ideas and themes, one of its most crucial aspects is a focus on the sense of *disembodied presence,* a feeling that pervades the viewer gradually as the film progresses, whether it happens to be a conscious or an unconscious feeling.[1] Just as astronauts who are completely surrounded by a mechanical environment feel less and less connected to the natural world beyond the spaceship walls and console panels and perhaps even less connected to their own natural bodies, so does the film viewer feel more and more disconnected from the familiar world to which his fives senses are normally attuned.

Kubrick presents us with different worlds, whether prehistoric or futuristic, worlds that are not like anything we have ever experienced. And the ways in which the director constantly changes the perspective of the audience throughout the film—changes that are occasioned through dramatically varying camera angles and movements as well as by vast and sudden jumps in narrative development—lead the viewer further and further from any sense that he is rooted in a localized, fixed, and continuous vehicle of perception, something akin to a material body. The more that our perspective or sense of physical identity changes radically, the more disconnected and detached we feel from a natural world that occasions our overall feeling of stability, continuity, and coherence.

For example, in the Dawn of Man sequence in the earliest part of the film, the prehistoric landscape is presented to us in a series of very specific views from fixed standpoints, passing one into another like some photographic slideshow, with one shot fading into view and then fading out as another is about to emerge, like acts of instantaneous creation out of nothing (echoing the book of Genesis's account of God's *creatio ex nihilo*). But from whose perspective are we supposed to absorb this landscape? Is it the viewer's standpoint alone, without any connection to a particular character—something akin to a God's eye point of view? Indeed, throughout most of the film, Kubrick seems to give us a history of the development of human intelligence from the perspective of some detached and impersonal cosmic spectator, and especially one that is not bound by the limitations and fixity that typically define the viewpoint of a finite and embodied human being, of an individual who is located in some given place and moment of the here and now.

Conventional films tend to introduce specific characters from whose viewpoints we process the rest of the visual and aural information that is given to us. Or we at least become associated emotionally with the perspectives of different characters as the film narratives progress in their individual ways. In John Ford's *The Searchers*, to take but one example—and one that also focuses on an odyssey through space and time, and that likewise makes use of Monument Valley, Utah—we are initially given a time and a place ("Texas 1868") just as we are given a more general reference to a specific initial setting in *2001* ("The Dawn of Man"). As Ford's film begins, we see a door swing backward, opening onto a brilliant western landscape. We see this from the standpoint of a woman (Aunt Martha, played by Dorothy Jordan) who is momentarily unknown and anonymous but who then enters the frame by exiting through the door and away from the camera, out of the blackness surrounding the doorway. She is initially silhouetted by the darkness of the house but then becomes clearly visible as she moves through the doorway and onto the porch beyond it. The camera zooms toward the fragments of bright blue sky and fiery orange desert that indicate the landscape beyond the passageway. The camera then moves through the opened doorway, following the woman as she emerges onto the porch to look for an approaching figure that she apparently spied from a window a few moments before this.

In Ford's opening sequence, by way of contrast with Kubrick's strategy in *2001*, the camera angles and motion are quickly tied very specifically to a physical context in which we now find ourselves, since we have met—at least superficially—the character to whom our perspective was originally conjoined. And as we quickly meet new characters (Martha's family), we view them either from the perspective of a character whom we have already met or from the standpoint of an anonymous observer who is nonetheless already familiar with the general situation (i.e., that of characters we know as well as new ones arriving into the scene). The viewer feels bodily located in the scene, since there is a sense of the fixed presence and continuous narrative identities of the characters, along with a limited selection of viewpoints due to the limited number of characters to whom we turn our attention—just as we have a limited number of possible perspectives from which to view an object in a room. The limitation on the number of possible viewpoints is given by the very fact that we perceive things via our finite bodies, from a definite and specific physical location at any given moment, and the continuity of our material existence affords a certain unity and continuity to our world of manifold mental and perceptual phenomena. Likewise, as we gain empathy or merely relationships with characters on a film screen, characters whose personalities unfold within a specific world that is signified by a specific narrative, we use our reason and imagination to align our intellects and emotions with particular characters who are grounded in fixed, localizable bodies that occasion a limited rather than infinite range of possible viewpoints.

In most of *2001*, on the other hand, Kubrick cleverly uses camera angle, camera motion, montage, and mise-en-scène to deny or severely limit any feeling of being rooted in a familiar world that affords physical location in terms of fixity and continuity. Unlike Ford's conventional filmmaking as exemplified above, where perspectives are tied to specific characters or at least to very specific locations and contexts, Kubrick's approach gives the audience a sense of being unrooted or physically disconnected. This is done by impelling the viewer to constantly question, at least in an implicit or even subconscious way, From whose perspective am I watching this image or character?

Now this is not to say that there is never a point during the film when the viewpoint of Kubrick's camera coincides with the viewpoint of a specific character. After all, as but one example, occurring just before the intermission or midpoint of *2001*, the viewer sees the astronauts' lips move from the fixed perspective of HAL the computer, who registers their conversation by reading their lips while they are huddled in the space pod. And when astronaut David Bowman (Keir Dullea) attempts later in the film to terminate HAL's overall operating system, Kubrick turns for the first time to a subjective use of the handheld camera, a technique that reveals Bowman's unique personal perspective and emotional context as he moves through the spaceship to dismantle HAL. And there are also certainly moments in the film when the viewer feels like an anonymous but engaged voyeur who is taking in a very specific situation from a series of particular locations within the context at hand.

But by and large, Kubrick shoots and edits *2001* in such a way that the viewer feels more disconnected than connected, more unrooted than rooted, and especially in a way that denies the sense of being physically anchored to a particular figure within the given situations on screen. Nor are there any characters in the film with whom we feel any solid emotional connection or identification. In fact, as many viewers of the film have indicated, the computer HAL somehow seems more human and empathetic at times than do the two astronauts, Bowman and Frank Poole (Gary Lockwood). And the minimization of human dialogue in the film (approximately 40 minutes of dialogue out of 141 minutes of running time) does not help matters in terms of identifying with certain characters.

Kubrick takes the audience on an audiovisual journey in which the personal sense of a fixed physical location or identity is fleeting because the viewer is almost always on the move, so to speak, across vast stretches of space and time. In this sense, cinematic form matches narrative content in a rather harmonious way. But Kubrick also implicitly forces us to consider the possibilities of an experience in which the role of the natural body—as the filter of one's individualized experiences and point of orientation for one's physical existence—is no longer primary. This is especially the case when we consider the fact that our technology has increasingly gained the capacity of delivering a more indirect and more generalized world, one in which our five senses play a minimized and mostly passive role. Today we have more possibilities of living in a desensitized and machine-molded environment from which we immediately feel detached and disconnected.

You Are Now Disconnected

As indicated earlier, our bodies give us our specific points of orientation at various moments and therefore occasion our specific perspectives on things around us. Our physical existence gives us direct experience of the world and of ourselves. And occasionally we feel the need to reaffirm the primary role of our physicality and our capacity for direct experience. We do this through athletic competition, hands-on training, risky adventure, and on-site sightseeing, to name but a few such activities. In *2001* this need for reaffirming the body's immediate presence and potential is emphasized in an intellectual manner since, as the film develops, the capacity of the characters for active and direct experience of the natural world becomes reduced and even nullified. This is analogous to a household item's becoming more noticeable once it is broken or goes missing.

A direct and active form of experiencing the world through one's physical senses is evident in the prefatory Dawn of Man sequence, not merely in terms of the ape creatures' overall reliance upon their bodies rather than their minds but more specifically in the scene where the apes express a need to touch the mysterious monolith. We see a group of ape-men huddled around the towering black rectangle, touching it as if to reconfirm that it is indeed there or that it is not going to move or that it does not pose a physical threat. Here we see the importance of direct bodily sensation—compare Dr. Heywood Floyd (William Sylvester) and his crew's attempt to touch the monolith later in the film, where the need for tactile contact is repeated, but, attired in their space suits, they are denied direct contact. All that Floyd can actually sense when touching the monolith is the feel of his own clothing and spacesuit.

A sense of disembodiment—of experiencing things in the world without a sense of fixed and continuous physical identity and without a direct and active use of the five senses—is also instilled in the viewer during the famous Waltz of the Spaceships sequence. This part of the film occurs immediately after Kubrick's legendary jump cut from the Moon Watcher's bone tool thrown into the air to (eons later) a similarly shaped and similarly colored spaceship flying through the extraterrestrial darkness, moving to Johann Strauss's "The Blue Danube." From whose perspective do we see the different ships in flight? Evidently from the standpoint of a detached and dislocated observer who hovers in space like a mind without a body, or at least a mind with a body of no fixed physical location. And even during the segment of the film in which we, identifying with the camera's viewpoint, feel as though we are soaring across a given distance of space with our physical identity intact, Kubrick immediately cuts to a new perspective and a new situation and a new object, as if we are switching identities with a series of anonymous observers. Now it is true that cutting to a new perspective or changing our identification with characters happens many times in films. After all, think of chase sequences where we see the chase from the viewpoints of both pursuer and pursued as well as from the perspective of an outside observer, to take one example. But in 2001, given the thematic content of the film, Kubrick expands the cinematic possibilities for the audience to experience repeatedly a seeming escape from the fixity and finitude of the individual body.

The audience's sense of detachment and even disembodiment increases as Kubrick intercuts the exterior scenes of white spaceships soaring against the canvas of star-flecked space with interior scenes of Dr. Floyd asleep in the vehicle that transports him to the wheeling space station. There he will embark on the remainder of his journey to the lunar station Clavius. As if we were not disoriented enough, though in a surreal and meditative (rather than disturbing) sense of dislocation, we are impelled to ask ourselves, What sense of viewer identity allows us to jump back and forth between interior and exterior scenes amid the emptiness of outer space? After all, as yet we do not really know who Dr. Floyd is, and so our association with his perspective is minimally engaging at best. To make matters more complicated, he is presented to us while he is asleep, and in addition he has no direct relation to our own freewheeling view of the ships waltzing through space before we actually meet him. At best we must align ourselves with the context-free perspective of a nonlocalized, omnipresent eyeball that sees almost everything from almost every perspective possible. In fact, such an eyeball serves as a recurring symbol throughout the film: HAL's red-lit eyeball that appears in multiple locations throughout the spaceship, astronaut Bowman's eyeball that stares out the window of his space pod as he witnesses visual leaps across space and time, and the metaphorical eyeball of Bowman's space pod window itself as it provides the lone surviving astronaut with a glimpse of the unfolding spectacle that is the Star Gate.

In the Waltz of the Spaceships scene, there are other hints of detached or disembodied presence, mainly in the form of zero gravity: Dr. Floyd's pen floats and twirls in the space of the rocket cabin as he sleeps, free of Earth's pull, in contrast to the heavily constrained bodily motion of the stewardess who returns the pen to Floyd's pocket. The sense of being disoriented in space is also implied subtly by Floyd's conversation with the Russian scientists before his flight to the moon. Floyd asks them, "Where are you all off to? Up or down?" The male scientist smiles and points down. But this is a silly question and an equally ridiculous gesture in response when one ponders the situation: up/down and left/right make no sense in space, since one must have not only a fixed point of bodily location but also some absolute reference such as Earth to give some context to any indication of directionality. Since these characters are in a space station orbiting somewhere between Earth and the moon, the distinction between up and down makes sense only by returning to an Earthbound or geocentric perspective. In outer space, there is no real meaning to such terms except in an obviously relative manner, given that there are also no absolute points of reference. Is the moon up or down when we realize that it orbits Earth continuously, and from whose perspective? Is Mars east or west of Earth when we realize that the planets orbit the sun at different distances and rates of speed?

On Floyd's trip from the space station to the moon so that he can investigate the recent discovery of the lunar monolith, there are further signs that Kubrick attempts to instill a feeling of disconnection and dislocation in the viewer. There is the reminder of zero gravity once again as we watch a stewardess deliver a tray of food to the pilots by walking in a circle upside down (at least from the fixed camera's point of view) so that she can enter the cockpit in a manner that is oriented to that new location in a different part of the ship, a new physical context that is given its own sense of directionality through the pilots' personal orientation. And so as to emphasize the radical change in directionality that is being offered here, Kubrick rotates the camera 180 degrees as the stewardess enters the cockpit and greets the pilots.

In conventional films like Ford's *The Searchers,* as mentioned earlier, the viewer certainly switches from the view of one character to another, or from one location to another, but there is an underlying sense of narrative context and physical continuity that aligns us with one or another of the characters in question. When such continuity is disrupted, as in a more unconventional film that switches alignment with characters in a radical manner (think here of a Robert Altman film like *Nashville* or Quentin Tarantino's *Pulp Fiction* or David Lynch's *Lost Highway,* for example), then a form of narrative dislocation (rather than merely spatial or temporal dislocation) is occasioned. Kubrick's method of dislocation in *2001* is more radical, however, in that the camera is often not associated with any characters or community of characters for very long, and the sense of space and time is already stretched to unconventional, cosmic lengths by the very plot and setting of the film.

A feeling of desensitized existence is also evoked when objects are denoted in indirect or denatured (e.g., symbolic) ways. For example, during Floyd's trip to the moon after conversing with the Russian scientists in the space station, he consumes a meal that is depicted by pictorial symbols for the different types of food that are contained in the box. This reveals an indirect presentation of the food itself, unlike the graphic depiction of food that we find in the Dawn of Man sequence after the ape-man tears apart the flesh of a tapir once he has discovered the use of the bone tool as a weapon. When objects have become little more than pictures and symbols, we are already at a major remove from physical or natural reality. And as if to emphasize even further the rather unreal or artificial nature of the depiction of Floyd's food tray, Kubrick shows the tray accidentally floating in the air as Floyd speaks with a fellow passenger.

The theme of disconnection and detachment becomes even more pronounced in a narrative manner as the film progresses. We witness Poole's death while he is disconnected from the mother ship, Bowman's disconnecting of HAL, and then Bowman's eventual separation from the mother ship as he is pulled into the Star Gate through the monolithic portal. One would like to say that Bowman's departure from the computer-governed *Discovery* and his subsequent experience of the Star Gate themselves constitute a return to direct and active perception, a return to one's rootedness in the senses of the physical body. However, Bowman can experience the passing phenomena of the Star Gate only by glimpsing them through the glass of his space helmet, through the window of his space pod. Bowman can be said to have direct and immediate contact with the world of the Star Gate only once he has exited his space pod and space suit, and this occurs only once he has become identified with the rapidly aging man in the Louis XVI–style hospital room, not to mention with the Star Child whose direct gaze at the audience concludes the film. And since these concluding scenes are highly surreal and symbolic, any chance of the viewer's recovery of some familiar sense of a stable physical identity at the end has been vanquished once and for all.

Philosophical Implications

And so Kubrick's *2001: A Space Odyssey* forces the thoughtful viewer to consider the implications as well as the dangers of going beyond our conventional conceptions of what it means to be human. This consideration has recurred throughout the history of philosophy.

For example, Martin Heidegger, the leading proponent of hermeneutic phenomenology, a branch of contemporary philosophy, emphasizes the fact that human knowledge—indeed human existence in general (what he calls *Dasein*, or "being-there")—is meaningful only in terms of finite contexts. These are contexts in which objects or events gain value or significance through their involvement in purposive relationships that are established by a self or subject that is bound by spatial and, more essentially, temporal limits. In making this point, Heidegger reacted against the earlier attempts of his mentor, Edmund Husserl, to secure a transcendental or unconditional standpoint from which conscious acts and relations could be constituted and made transparent. Heidegger came to realize that the idea of some standpoint of pure consciousness or transcendental subjectivity (as presupposed earlier in the tradition by René Descartes and then by Immanuel Kant and Husserl) is ultimately illusory and devoid of meaning. Only by being embedded within limited yet value-laden situations of the present moment, shaped by the past and projected into the future, could the human subject (as a form of being-there-and-then) possess meaning and value. In fact, Heidegger even rejects the notion of subjectness or subjectivity in general, since this idea implies not only some absolute division between self and object but also something that exists on its own, apart from various situations and contexts.

To think that one can operate at a transcendental or context-free level of cognition or from a desituated standpoint, as Kubrick's film invites us to suppose, is to deny our very embodiment (being-in-a-body, so to speak) and historicity (being-in-time). Our horizons of creative possibility are engendered, in fact, by our context dependency. Those who pretend to locate themselves somehow beyond the borders of their present life-situation are left only with nothing in particular—an absence of meaning and value. Simply put, according to the Heideggerian viewpoint, there are no absolute or Archimedean standpoints for beings such as ourselves. Our experience and knowledge are always dependent upon our physical embodiment and upon background contexts that are not immediate objects of perception or cognition. To experience or to know something, according to Heidegger's famous work *Sein und Zeit* (*Being and Time*), we must always already be located in space and time, with a given orientation and perspective, which means that we can become aware of the world only once we are actually engaged in specific and meaningful situations.

The nihilistic or life-negating dangers of any attitude that downplays the importance of the human body and its limitations are amplified when we ponder the type of detachment or disembodied feeling that an overemphatic use of technology can generate. This topic is treated in an illuminating way, for example, by Husserl and Heidegger scholar Hubert Dreyfus in his book *On the Internet*. Dreyfus points out that an excessive dependence upon the Internet (along with the growing possibilities of virtual reality) creates an increasing sense of detachment that removes the Net user from value-laden, risk-involving situations in which the knower is intimately engaged in a given inquiry or project.

The author uses distance learning, or remote education, as a case in point. By examining different levels of the instruction process, Dreyfus argues convincingly that, although online learners may easily become novices or advanced beginners through remote, technologically delivered teaching, they cannot proceed far beyond those two initial stages of education without accepting a special kind of involvement.[2] This involvement demands that the student engage in concrete situations, take risks, and create personal perspectives on the subject matter at hand, but these activities require being physically present in a specific context, with a specific bodily point of orientation. It is precisely the type of fixed bodily orientation that Kubrick denies us as film viewers throughout major segments of *2001*.

To frame his point about the existential, psychological, and moral dangers of technology in general and the Internet in particular, Dreyfus refers at the outset of his inquiry to an implicit debate between the ancient Greek philosopher Plato and the modern German thinker Friedrich Nietzsche.

Plato, via his character Socrates, suggests a form of detached and nonphysical existence in which the immaterial mind and intangible soul are given priority over their temporary vehicle, the physical body, along with its corresponding senses and desires. In fact, for Plato's Socrates, the death of the body is in many ways the full awakening of our genuine intelligence. Only when the mind departs from its physical confinement can it truly and completely grasp those eternal Ideas or Forms that are the universal patterns of reality. The best that we can do as living Earthbound creatures, according to such a view, is to attempt a concentrated but inevitably partial withdrawal of our minds from the desires, distractions, and dolor of our physical embodiment.

Along these lines, Plato emphasizes the need for philosophical reason to distance itself as far as possible from the subjective instincts and desires of the body in order to attain the ideal of objective knowledge. As we learn from Plato's dialogue *Phaedo*, for example, philosophy is associated (at least figuratively) with the disembodiment of death in that genuine philosophers must strive to purify their souls as much as possible, rising above the particular needs and desires of the situated, individual body. Dreyfus quotes from the *Phaedo* in his introduction, citing Socrates' telling statement: "In despising the body and avoiding it, and endeavoring to become independent—the philosopher's soul is ahead of all the rest."[3] So Plato, according to this interpretation, advocates a brand of intellectual and even spiritual detachment that denies the concrete particularities of life rather than affirming them.

Nietzsche, on the other hand, rejects Plato's view and emphasizes the importance of one's body as it is situated in the spatiotemporal tapestry of the natural world. For Nietzsche, it is clear that any attempt to transcend or deny our earthly, bodily finitude results first in the erection of abstract, illusory ideals and then in a detached, life-negating existence. Dreyfus quotes from Nietzsche's masterwork *Thus Spoke Zarathustra,* where Zarathustra proclaims in a chapter titled "On the Despisers of the Body," "I shall not go your way, O despisers of the body! You are no bridge to the overman! . . . Behind your thoughts and feelings, my brother, there stands a mighty ruler, an unknown sage—whose name is self. In your body he dwells; he is your body."[4]

The master morality and creative individuality of Nietzsche's ideal personality type, the Übermensch (superman), does not involve a denial of earthly existence but rather seeks to affirm and celebrate the world of the here and now through acts of self-overcoming. For Nietzsche, self-transcendence is not about the rejection of our situated, bodily, and mortal nature. Rather, we are able to transcend particular limitations, obstacles, and perspectives because of the very fact that we are situated, embodied, and mortal creatures in the first place. Our bodily and earthly finitude, in other words, makes possible any act of life enhancement, since it is our physical embodiment that affords us the specific contexts or situations by which we can define ourselves in a dynamic and self-transforming manner.[5]

Dreyfus makes the case that our world is increasingly governed by machines and cluttered by artificial substitutes for the natural environment. Coping with such a world can lead to a detached and life-negating form of existence, one that decreases the active role of the body and that severely limits our choice over physical perspectives. You might think here of the commitment-free, risk-free, hedonistic Web surfer, someone who sits in front of a computer screen and bounces mentally and whimsically from one Web site to another, without any ultimate passion or active engagement. Such a form of obsessive detachment from the natural world and from the experience of one's own body can lead to nihilism, the belief that nothing matters, in the undermining of our commitments and concerns. We would become little more than passive and neutral spectators, gliding from one Internet portal to the next, and our daily lives would follow suit.

Kubrick's *2001* impels us to ponder the moral, psychological, existential, and even spiritual dangers of our drive to transcend our general state of humanity, particularly the problems that may arise if we ignore the requirements that our physical embodiment imposes upon us. But *2001* is also a film that evokes questions about the possibilities and implications of artificial intelligence, a form of intelligence that is exemplified by the mechanical character of HAL in the film. How humanlike might a computer actually become as technological progress continues at a rapid rate? After all, not only is HAL intelligent in the sense of processing information quickly and executing commands, but it even seems to display emotions, a power that actually endangers the human crew of the spaceship when HAL's personal stability, so to speak, comes into question.

In a chapter on artificial intelligence in *On the Internet,* Dreyfus suggests that one reason why our current scientists and researchers have not yet developed a computer that can search for meaning in the way that humans can is the simple fact that machines do not possess the kinds of bodies that we do. Computers and computerized robots do not experience the kinds of sensations and emotions—like those of curiosity and risk and sacrifice and satisfaction—that make us respond to situations in very human ways. Without the range of bodily options and orientations that we as human beings encounter from moment to moment, simply because we are defined by our physical situations and the choices that they impose, machines cannot hope to imitate the full depth and breadth of human intelligence. Thus far, the only computers that have been able to sense and experience—indeed, feel—the world as humans do are those that we see in films.

Notes

1. The term *disembodied presence* is taken from Hubert L. Dreyfus, *On the Internet* (London and New York: Routledge, 2001).

2. Ibid., 35.

3. Ibid., 5.

4. Nietzsche quoted in ibid.

5. As Dreyfus summarizes, "Nietzsche thought that the most important thing about human beings was not their intellectual capacities but the emotional and intuitive capacities of their body" (*On the Internet,* 6).

Terminator-Fear and the Paradox of Fiction

Jason Holt

POSSIBLE RESPONSE[S]:
YES/NO
OR WHAT?
GO AWAY
PLEASE COME BACK LATER
. . .

—The Terminator's language processor (heads-up display)

Some of the most vividly unnerving scenes in *The Terminator* (James Cameron, 1984) are those that present the Terminator's point of view, giving us a sense of what it would be like to be the Terminator, to see the world as it does, to have not only artificial intelligence but also, more disturbingly, artificial consciousness. The judicious use of the subjective camera is an especially effective technique when appropriately modified to evoke alien perspectives, those radically unlike our own. The Terminator's visual field is infrared, with heads-up displays for attentional shift and focus, information processing of different kinds, decision-making menus, and action-guiding schematics. Although in some sense we can never know what it is like to be a creature whose consciousness is radically different from our own, *The Terminator* gives us the imaginative wherewithal to grasp what an artificial consciousness might be like, more vividly and effectively perhaps than other AI-heavy films, such as *2001: A Space Odyssey* (Stanley Kubrick, 1968) and *Blade Runner* (Ridley Scott, 1982), which neglect or marginalize the first-"person" perspective on artificial consciousness, especially as it is likely to be something radically unlike our own.[1]

The basic premise of *The Terminator* is as follows: from a postapocalyptic future in which humans are on the brink of winning a long-standing war with rogue machines, a Terminator (Arnold Schwarzenegger) is sent back in time to kill Sarah Connor (Linda Hamilton) before the birth of her son John, the leader of the human revolt and sine qua non of its success; to protect his mother and save himself from "retroactive abortion," John sends back his brother in arms and not-yet father, Kyle Reese (Michael Biehn).[2]

Time travel and artificial intelligence, not to mention a healthy (or un-healthy) dose of technophobia and technophilia, are science fiction staples, and *The Terminator* blends them well in its cinematic style (viz. the nightclub aesthetic of tech noir). The received wisdom is that both artificial intelligence and time travel are possible, the latter at least theoretically, and though aware-ness of this should inform our reactions to science fiction scenarios, the more central issue is the emotional reactions themselves.[3] I suggest that we find the Terminator's alien, in many ways superior, artificial consciousness to be at least as unsettling as the physical threat the Terminator poses. This is brought home more than anything else by the first-"person" Terminator-perspective shots. When the Terminator chases Sarah and Reese down a dark alley, its unimpeded infrared visual display calls up a crosshair scope to track them, focusing the Terminator's visual attention and mortal intention alike. At the police station shoot 'em up, its vision unimpaired in the blackout, the Terminator perceives and acts in what appears to be slow motion. When it digitizes and stores information, such as the address of Sarah's mother's cabin, we know that the information will not be forgotten. When it climbs into the cab of an eighteen-wheeler, it calls up a gearshift diagram that instantly en-ables it to drive a double clutch. Scary stuff, this: this alien, capable, ruthless consciousness. What makes the Terminator's physical threat so horrific is the notion that what drives the physically superior, lethal machine is, in relevant though not all respects, a superior, more efficient mind, a consciousness with which we simply cannot connect. As Reese says, "It can't be bargained with, it can't be reasoned with, it doesn't feel pity, or remorse, or fear, and it absolutely will not stop, *ever*, until you are dead."

Whereas the Terminator does not feel pity or fear, we do, not only in everyday life but also in our encounters with fiction. Our emotional reac-tions to artificial intelligence and its fictional depiction in *The Terminator* will serve as focal points in this essay for examining what is often called the paradox of fiction. In a nutshell, why do we respond emotionally to fictions, things that we know do not exist? Why does the Terminator frighten us?

After explaining the paradox in some detail in the next section, I will argue for a contextual solution, one motivated by a divide-and-conquer strategy.[4] What makes *The Terminator* a particularly useful paradigm case is that, among various quality science fiction films, it stands perhaps unsurpassed in generating a particularly significant range of certain emotions (fear, pity, admiration, desire). When this range is examined vis-à-vis *The Terminator*, a key assumption underlying the paradox (specifically that emotions are homogeneous in paradox-relevant respects) is exposed as problematic.

Here is a thumbnail sketch of what follows. Sometimes, owing to context or the kind of emotion in question, our response to fiction is not genuinely but rather ersatz or quasi emotional. When the emotional response is genuine, however, we need not believe that the fictional object eliciting it somehow really exists. We might well experience real fear when the Terminator is on screen, even if the pity-like response we have toward its victims is ersatz. Such responses can be rational, furthermore, when the possibilities the fiction presents are in palpable measure plausible. Increased awareness of the possibility of artificial intelligence makes the scenario presented by *The Terminator* seem less like science fiction and more like science future. There is room to speculate, in offering an alternative solution, that since fictional entities can serve to focus abstract, emotionally relevant concerns, they not only furnish us with but also in some sense are, and not to the detriment of our reason, objects for these emotions.

The Paradox, Suspending Disbelief, and Make-Believe

Without reflecting on it too much, we might assume that our emotional responses to fiction, including movies like *The Terminator*, are unproblematic. Getting caught up in a good story often means, among other things, being moved by fictional characters, by what happens to them, by what they do. So the Terminator makes us feel fear, Sarah pity, and Reese admiration. What could be more straightforward? On reflection, however, there is something odd, even paradoxical, about our emotional responses to fiction. The paradox is generated by the following three individually probable, almost axiomatic, but mutually inconsistent propositions:

(1) Readers or audiences often experience emotions such as fear, pity, desire, and admiration towards objects they know to be fictional, e.g., fictional characters.

(2) A necessary condition for experiencing emotions such as fear, pity, desire, etc., is that those experiencing them believe the objects of the emotions to exist.

(3) Readers or audiences who know that the objects are fictional do not believe that these objects exist.[5]

In sum, how is it that we feel emotions toward things we know do not exist, when such feelings apparently depend on believing they actually do exist? How is it, for instance, that we can feel fear toward the Terminator when, as far as we know, there is no such thing? Formally speaking, there are eight ways of handling the problem, not all of them equally plausible. We might try to reject any one of the three propositions (three), any pair of the three (three), all three (one), or none (one), in the last case concluding unpalatably that something in our engagement with fiction is fundamentally irrational, since the triad implies that an emotional response to fiction involves the belief that a fictional entity, in the same sense, both does and does not exist. I will focus on the first and last maneuvers here, that is, attempts to reject a single proposition and, barring that, acceptance of the verdict that emotional engagement with fiction is irrational.[6] This will help to motivate a contextual solution that limits the scope of propositions 1 and 2 without rejecting either.

A solution with traditional flavor but without much current support is to reject proposition 3, on the grounds that when audiences are truly engaged in a fiction, they come to believe, in a way, or half believe that the fictional characters and goings-on are real. When we fear the Terminator, we come somehow to believe, or half believe, that it is real. This should recall Coleridge's famous "willing suspension of disbelief," although the positions are not the same. Whereas naive audiences might briefly mistake fiction for fact, "sophisticated audiences, to whom the paradox is addressed, do not come to believe, or even to half-believe, that fictional characters are real people, though such characters might *seem* quite real."[7] However real the Terminator seems, no one really believes or half believes it to be real. No wonder that such a solution strikes one as, to say the least, implausible. It is. But we should take pains not to conflate this proposal, rejecting proposition 3, with the notion of suspending disbelief.[8] Since the disbelief in something fictional is the belief that it does not exist, suspending that belief does not imply the belief that it does exist, just as a suspended atheism fails to imply theism. The disbelief simply fails to play the active role in one's mental economy that it will in other contexts. One simply goes offline. To get caught up in *The Terminator*, it might help not to worry about the metaphysical status of the things I seem to see. That would distract from the entertainment. But again, this does not mean that I believe what I see. So Coleridge's view is a rejection not of proposition 3 but of proposition 2, a type of solution to which we will return in later sections.

Audiences typically do not believe that fictional objects are real (except, insofar as the stories exist, as fictions; the Terminator is not real, but *The Terminator* is). It may also be that our emotional responses to fiction, in some sense, are not real either. In other words, the culprit may be proposition 1. To clarify this, on Kendall Walton's account of artistic representation and engagement as a kind of make-believe, we do not have bona fide emotional responses to fiction, even if those responses feel like the genuine articles.[9] At most we experience quasi fear or quasi pity, ersatz admiration. Whereas real emotions dispose us to take action, fleeing the fearsome, comforting the pitiable, and so on, our affectively charged responses to fiction do not, in any ordinary way, so dispose us. When the Terminator comes on screen, no one flees the theater or switches off his or her DVD or VHS player. Still, knowing that the fiction is a fiction, in concert with bona fide emotional response, would seemingly suffice to explain the dispositional lack. Critics observe that the make-believe account requires both that audiences be systematically mistaken about their emotional states in encounters with fiction, unable to distinguish, say, between real and fake pity, and that, for instance, horrific scenes cannot yield genuine fright, only an undetectable knockoff version.[10] The critics are certainly onto something. The outright rejection of proposition 1 is probably not viable as a general solution to the paradox. By the same token, we should not be too quick to dismiss its potential use in a curtailed and slightly modified form, as we shall see.

Divide and Conquer I: Context and Emotional Kinds

Obviously it is not true that everyone always responds emotionally to fiction in encounters with it. Watching *The Terminator*, a person need not be moved to fear the Terminator, pity Sarah, or admire Reese. For a number of reasons a fiction, even a well-done fiction, may leave us cold or, worse, turn us off. Such cases are not central to the paradox, however, which does not concern the failure to respond emotionally or such emotional responses as we might have to the film qua film—liking it, for example, or disliking it—but rather responses to fictional elements in it, the difference between disliking *The Terminator* and disliking the Terminator.[11] It is the standard cases that count, those in which we are engaged with and actively appreciate a fiction. However, appreciatively engaging a fiction does not by itself imply emotional response to any, much less all, of its fictional elements. Assuming that we do respond emotionally to "fictionalia" in a broad range of standard cases, this is perfectly compatible with there being a broad, complementary class of standard cases in which we fail to so respond, in which we do imagine or pretend that we are having the relevant emotions. If we grant that audiences often feel real pity for Sarah or fear of the Terminator, it is no less plausible that audiences often, in some sense, merely imagine doing so, approximating the mind-set, perhaps as a way of facilitating aesthetic pleasure in the absence of robust emotional engagement. The key difference between this perspective and Walton's, aside from its being offered as only a piece of the completed puzzle, is that audiences clearly know the difference between feeling and faking it, between genuine emotion and make-believe as considered here. More important, we leave the door open for plenty of real emotional responses in a broad range of standard cases.

If it is at all plausible to suppose that our emotional responses to fiction can be, by turns, bona fide and discernibly make-believe, this is likely not only because of the different contexts of engagement but also because of differences in emotional kind, the type of emotion in question being more significant than is usually supposed. As the very formulation of the paradox suggests, proposition 1 especially, there is a tendency in the literature to treat various candidate emotional responses to fiction the same, as if there were no paradox-relevant differences among what seem to be, in the case of watching *The Terminator,* my pity for Sarah, admiration for Reese, fear of the Terminator, and desire for Ginger (Sarah's roommate). Emotions are assumed to be homogeneous in paradox-relevant ways, but this presumption of affective unity is ultimately untenable. Some emotions, like fear and desire, are lower, more basic, than others, are experienced by a wide variety of animals, require little if any intellectual involvement or cultural understanding, and are rooted in evolutionarily primitive parts of the brain, specifically the hypothalamus and limbic system. Other emotions, such as admiration and pity, are less basic, higher, as they are experienced by comparatively few species, require more cognitive sophistication and cultural participation, and are elaborated from the reptilian brain into the cerebral cortex. Because of the greater cortical involvement and, for that matter, commitment required by higher, nonbasic emotions, it is more than plausible to suppose that it is in connection with them, and not with the more basic emotions, that make-believe responses to fiction predominate. Maybe we cannot often or ever literally pity Sarah or admire Reese, as pity and admiration more intuitively seem to require an existential commitment, a belief that the objects of the emotions exist, far more than do feelings of fear and desire. It certainly seems to me that I do not literally pity Sarah, although by contrast I do acknowledge a mild, somewhat excited fear when I see the Terminator on screen, scored by an unnerving four-beat-cycle artificial heartbeat. Note that the criticism of the make-believe view, discussed above, that it is unrealistic to say that we never get genuine emotions vis-à-vis known fictions, is most persuasive when considering the lower, basic emotions, and not nearly so otherwise.

If the preceding discussion is on the right track, then in many contexts, and for what I am calling higher emotions typically, we have reason to reject proposition 1 as requiring that audiences often experience genuine pity, admiration, and the like toward objects they know to be fictional. We also have reason, in many other contexts, and for what I am calling lower emotions, to maintain proposition 1 as requiring that audiences often experience genuine fear, desire, and the like, *toward*—the preposition will become all-important later on—objects they know to be fictional. We must seek the solution for lower emotions elsewhere, and this is next.

Divide and Conquer II: No Object Required

So far we have half of a contextual solution to the paradox. While we might sometimes think we feel genuine pity for Sarah or admiration for Reese, we usually and knowingly do not, because of the existential presupposition of higher emotions. This leaves lower emotions, such as fear toward the Terminator, unaddressed, since the persuasiveness of a make-believe-style solution depends on their exclusion. With propositions 1 and 3 intact, to avoid the implication that engagement with fiction is fundamentally irrational, we must look to scotching proposition 2 as it stands, according to which having an emotion implies belief that the object exists. The rejection of 2 normally involves some species of so-called thought theory, according to which the fiction-to-emotion process, mediated by thought, may and often does occur without concomitant existential belief. The thought alone is sufficient. Fear is caused by thoughts about the Terminator, but it is not fear *of* the Terminator, the *of* phrase deflated to specify nothing more than nonrelational content.[12]

In other words, my fear *toward* the Terminator, caused by the film and mediated by thought, is mere Terminator-fear, literal fear, but not literal fear of the Terminator. If emotions can be felt without the existential commitment required by proposition 2, then the proposition is false.

We have already seen that this is a more persuasive view of lower emotions than it is of higher emotions. Thought theory probably cannot provide a complete solution on its own, for reasons that complement those evincing the same outcome for the make-believe approach. Where the make-believe approach succeeds, thought theory fails, and vice versa. Thought theory will prove at best a good account only of the lower emotions. Whereas the make-believe approach, dispensing with genuine emotion, need not posit objects of quasi emotion, especially when construed as discernible as such by their subjects, as I have urged, the thought theorist is obliged to explain away the apparent need. Indeed, the standard worry about thought theory is that fiction-generated emotions have no corresponding objects, that when I feel, say, Terminator-fear, there is nothing of which I am afraid, apart from, in some sense, the imaginary object, which I seemingly do not take to correspond to anything real.[13] Other than simply insisting that the imagined Terminator is sufficient for my fear independent of existential commitment, one might claim that although there is strictly nothing of which I am really afraid, I am nonetheless afraid of it, in the same way that while it is true that the ancient Egyptians worshipped Osiris, there is nothing of which it is true to say the Egyptians worshipped it.[14] But this reply neglects a crucial disanalogy: the Egyptians believed that Osiris existed; I do not believe the Terminator does. I cannot fear the Terminator in a sufficiently analogous way. There might be nothing real in either case, but the crucial thing is what the parties take to be real.

What we need at this point is independent reason to suppose that lower emotions like fear do not require objects, do not depend in other words on the subject's existential commitment. For starters, as Peter Lamarque observes, "that thoughts [alone] can have physiological effects is well recognized in the case of revulsion, embarrassment, or sexual arousal."[15] We might also observe that certain emotional states, like free-floating anxiety, seem not to require, and perhaps by definition cannot have, an object. As well, the notion that emotions are invariably caused by thoughts is too simplistic if not naive. Neurologically speaking, basic emotional response often precedes thought, occurring before existential considerations can enter in—hypothalamic and limbic first, and only then cortical—and somewhat recalcitrant to such afterward, with or without a presumed object. Seeing the Terminator on screen may, in the right circumstances, unavoidably elicit a fear response regardless of one's unshaken belief that it is not real. Not only is existential commitment unnecessary in these cases, so too, at first, is thought. Another, more general, example is fear of the future. The future does not yet actually exist, but prospective events and not yet real things can nonetheless be feared. One might fear artifacts like the Terminator being built one day.

To take an example closer to home, suppose I am awakened one night by a noise, and as a result I am agitated, fearful, and I speculate that someone may have broken in or that some other danger, some imminent harm, might be in the offing. In this state I might be genuinely frightened, but I do not necessarily believe, and I probably do not believe, that there is something of which to be so. What is implied here is not the belief that there *is* something of which to be afraid but instead the belief that there *might be* something of which to be afraid. The *might* makes all the difference. The possibility of there being something fearsome exists, but the possibly fearsome something need not. We must be careful to note here that the fear and the associated belief are distinct states of mind. It is not as though I fear the possibility of an intruder. Rather, I fear that there actually is an intruder. But I need not believe that an actual intruder is there. The most it seems I need to believe is that an actual intruder might be there. To fear the actuality, one need only believe the (presumably nontrivial) possibility. In fact, if I do believe that an intruder is there, I am likely not to fear it. What I am likely to fear instead is what the intruder will do, but has not yet done, to me or my property. So here again, while the fear represents actual states of affairs that are not yet actual, events that may or may not happen, I need not, and in some sense cannot, believe that what is feared is real.

Artificial Intel and Metaphysical Psych

It appears that we now have a neat contextual solution to the paradox of fiction. For higher emotions, such as pity and admiration, we limit the scope of proposition 1 to exclude such emotions. For the most part, we construe paradox-relevant prima facie cases, such as pitying Sarah, to reduce to quasi emotional states, with the twist that audiences can usually tell the quasi from the real. For lower emotions, such as fear, we limit the scope of proposition 2 to exclude such emotions, holding a modified thought theory, taking thought as sufficient sans presumed extant object to yield real emotion.

Possibilities will do. By such means we adopt the strengths of both theories and the weaknesses of neither, as we can ill afford to reject proposition 1 or 2 outright, and we avoid having to accept all three as originally formulated, which jointly imply that human engagement with fiction is fundamentally irrational. But the story does not end here, for while my Terminator-fear fails to imply that I believe that the Terminator both exists and does not exist, it might seem irrational for us ever to respond emotionally to fiction. If proposition 1 holds in a significant, though significantly reduced, range of cases, why should this be so?

Part of the answer might be suggested by the discussion of lower emotions in the section above. Such emotions are designed, and for good reason, to precede thought often enough. The mechanisms of emotion are biologically tailored to serve a number of fairly obvious functions in daily life, and though they frequently stir up psychological conflict and other sorts of trouble, they are pretty good at their job. They have a built-in practical rationality, and this is so even if, in certain biologically unusual domains such as fictional engagement, where makers of fiction play on such mechanisms, they systematically fail. So the mechanism is not irrational generally speaking, for failure in select domains can be counted as a by-product of what makes the mechanisms successful in domains of more immediate biological significance. But are the mechanisms then irrational within the domain of fictional engagement, where they systematically "fail" when fiction moves them, and consequently us, to experience real emotion? Arguably not. Perhaps real emotions reinforce the lessons gleaned from fiction in ways that quasi emotions simply cannot or cannot nearly as well. We might mention, too, Aristotle's familiar notion of catharsis—getting out the bad blood—or speculate, relatedly, that fiction exercises our emotional mechanisms, calibrating them for the real world, getting them set for life.

These considerations, while legitimate, might seem to miss the point. It is not the pragmatic, instrumental value of the mechanisms of emotion or the lower emotions they yield in engagement with fiction that matter but rather whether such emotions in such contexts are really warranted, justified, assuming it is appropriate to speak of the pure, abstract rationality of feelings in this way. Analogously, while the belief in the divine might be pragmatically vindicated—easing one's anxieties, furthering one's interest à la Pascal's wager—this does not mean the belief has any objectively rational justification. Maybe my Terminator-fear precedes thought and is to some extent resistant to thought's influence. Emotions may be cognitively impenetrable to a significant degree. But aside from the aesthetic pleasure facilitated by and incorporating my fear, what justification if any do I have for maintaining the emotion, for not trying to quell it to the extent that I can?

It might seem senseless to talk of the justification of emotion as analogous to the justification of belief. But a fear of Terminator-style tech noir seems, and is, far more appropriate in the context of a worldview that admits the possibility of artificial intelligence than in that of a worldview that does not admit it. A naturalist, who believes that everything in the universe, consciousness included, stems from physical states and processes, is apt to admit, almost by default, the possibility of artificial intelligence, of being able ultimately to build something like the Terminator. A dualist, by contrast, who conceives of the mind as a spiritual entity, one not physically determined, is liable, as of necessity, to deny the possibility. Assuming that both views are justified provisionally, if not when all the chips are down, the question of whether a *Terminator*-induced fear is rational ultimately depends on the subject's belief system, specifically the core of his or her conceptual framework, a question of metaphysical psychology. It is more rational for a naturalist to feel Terminator-fear than it is for a spiritual dualist.

At this point the discussion seems to have skirted a crucial if not directly germane issue: is artificial intelligence possible, really possible? The short answer is yes.[16] But this is not the relevant point here. The point is that experiencing Terminator-fear when watching *The Terminator* does not imply either unwarranted existential commitment or the subject's irrationality. Such emotions may be beyond critical censure anyway, even though there would still appear to be a marked difference between the naturalist's fear and the dualist's. The first would be more appropriate, more fitting, than the second. The upshot of this is that in entertaining genuine and genuinely fearsome possibilities relative to a subject's metaphysical psychology, there is nothing amiss in, and much to be gained by, the subject's responding accordingly.

Speculative Outro

Although I have offered a contextualist solution to the paradox of fiction, the search for a uniform solution is both psychologically tempting and perhaps theoretically desirable. One possibility that has not yet been raised is that we consider real-life counterparts to stand in as the objects of fictionally generated real emotions.[17] The standard example in much of the literature on the paradox is the pity we (allegedly) feel for Tolstoy's tragic heroine Anna

Karenina. On the standard counterparts interpretation, the pity we feel is real, but it is not really for Anna Karenina but rather for actual, real women who suffer similar misfortunes. This tack might seem less plausible in the case of science fiction, for the predicaments found in science fiction are unlike those in which any real person has been. Many people have become suicidal after being jilted by their lovers; no one has ever been hunted by an artificially conscious Terminator. Anna Karenina's shoes are often filled; Sarah Connor's, never. However, at the right level of abstraction, Sarah's predicament is quite familiar. She is overwhelmed by unrelenting imminent mortal threat, and many real people have been there. From another angle, although Sarah's specific predicament has no real-world counterparts as yet, it has an inordinate number of possible-world counterparts, and as argued above for the lower emotions, possibilities suffice. Perhaps they suffice for higher emotions too. Even if possible worlds exist in some sense, however, they do not exist in the actual way this world does. But no matter how we "modalize" the counterparts, this maneuver fails to account for the apparent particularity of our response to fictional characters: "We pity Anna Karenina herself, not just *women in Anna Karenina's predicament.*"[18] The problem, in other words, is that whatever might plausibly serve as the objects of real emotions generated by encounters with fiction are going to be too diffuse, too general, for the specificity of such responses. Such specificity might attach only to higher emotions, however, whereas the lower emotions are and ought to be more diffused. My Terminator-fear is clearly not of the particular Cyberdyne Systems Model 101 that targets Sarah but rather of tech noir generally. If this is so, then we simply return to the paradox and the contextual solution.

This return might be hasty, however, so let us speculate a little. Perhaps insights from the counterparts perspective, in abstracted or modalized form, can be blended with the concern motivating the specificity objection. Even in standard thought theory, there is a specific object, albeit unreal, toward which fiction-born and fiction-borne emotions alike reach. Our Terminator-fear is not strictly of, but is directed toward, the fictional object, thoughts of which inspire the fear. The Terminator, as depicted, gives us specificity, the counterparts plausibility. The key is to get the two together. One approach would be to liken the fictional object to a lens through which we metaphorically see the counterparts, just as views through microscopes depend as much on the lens power as on the nature of what is seen. Fear is directed toward the Terminator but does not stop there. In either case what is "seen," if not literally seen, can be presented and viewed in other ways, by other means, but in both types of engagement, we achieve a kind of situational union of the instrument of viewing and the thing or things viewed, even if the lens is less object than, in the right sense, objective. A somewhat different though related approach would be tantamount to identifying fictional objects with their counterparts. This might seem completely absurd, unless one considers that we can deploy a notion that unites the generality of a type (subsuming the counterparts) with the particularity of a token (a specific fictional object). I speak here of archetypes, which are particulars with universal punch, tokens that do not merely instantiate or even represent but rather embody their corresponding types. The Terminator, then, as fictional, may nonetheless be said to embody, as an archetype, and so in a sense *be*, the object of our fear; likewise for our pity of Sarah and our admiration of Reese. Although we do not thereby think that the Terminator actually exists as such, that is, without archetypal endowment, except as a fiction, in this sense the Terminator *is* tech noir, the object of our fear. Apart from presentation and representation, the fiction is the possibility.

What I have offered here, then, are two alternative solutions to the paradox of fiction, one straightforwardly contextual, the other, though highly speculative, more unified, an account that otherwise might be marshaled to defend thought theory, or the contextual view, from objections levied against the first but applying, it seems, to both. The speculative solution, I expect, will win few converts, as it is provocative but clearly needs much more fleshing out. I offer it to spur further creative thought on the paradox, not as a final solution. Perhaps, like many movies, books, and theories, it will itself prove, in this respect, at least a useful fiction.

Notes

Thanks to Steven M. Sanders for helpful comments on an earlier draft.

1. Thomas Nagel, "What Is It Like to Be a Bat?" *Philosophical Review* 83 (1974): 435–50.

2. The screenplay was written by James Cameron and Gale Anne Hurd, with additional dialogue by William Wisher Jr. The works of Harlan Ellison are acknowledged at the end of the film, although not specifically Ellison's "Soldier" and "Demon with a Glass Hand," teleplays for the original *Outer Limits* TV series.

3. If time travel is possible, then it must be possible to affect the past. But being able to affect the past does not obviously imply being able to change it. In the grand scheme, though not from Reese's timeline, he and Sarah conceived John before Reese

was sent back, before he himself was even conceived. If the past cannot be changed, then the Terminator's efforts must, of necessity, fail. However, the machines' metaphysics may descry a loophole neglected by these considerations. For more on time travel, see "Some Paradoxes of Time Travel in *The Terminator* and *12 Monkeys*" by William J. Devlin, in this volume.

4. This loosely follows the advice of Jerrold Levinson, *The Pleasures of Aesthetics* (Ithaca, NY: Cornell University Press, 1996), 303. Levinson suggests that different approaches to the paradox should be incorporated into a general solution to it. Importantly, however, a contextualist solution is not a general solution, although it does provide a framework for a variety of particular solutions. Peter Lamarque reads Levinson's position as broadly in line with Kendall Walton's, which will be discussed in the next section and which is decidedly not contextualist. Peter Lamarque, "Fiction," in *The Oxford Handbook of Aesthetics*, ed. Jerrold Levinson (New York: Oxford University Press, 2003), 388.

5. Lamarque, "Fiction," 386.

6. Colin Radford, "How Can We Be Moved by the Fate of Anna Karenina?" *Proceedings of the Aristotelian Society*, suppl. vol. 49 (1975): 67–80.

7. Lamarque, "Fiction," 387.

8. Contrast with ibid., in which the suspended disbelief view is closely associated, if not identified, with the rejection of proposition 3.

9. Kendall Walton, "Spelunking, Simulation, and Slime: On Being Moved by Fiction," in *Emotion and the Arts*, ed. Mette Hjort and Sue Laver (Oxford: Oxford University Press, 1997), 38.

10. Lamarque ("Fiction," 387) cites various critical responses to this effect, including his own *Fictional Points of View* (Ithaca, NY: Cornell University Press, 1996).

11. Perhaps we can avoid the paradox altogether by defending the idea that we never respond emotionally to the fictional elements in a work but only to a work's general capacity to entertain us and specific means of doing so; whereas the fictional elements may engage our imaginations unemotionally, it is the work and the engagement itself that serve as the objects of real emotions. As with the make-believe approach, this amounts to a rejection of proposition 1. But we are not forced, as on the make-believe account, to say that we have quasi emotions that cannot be distinguished from real ones. We simply find other objects for real emotions, the artworks themselves. This move is similar to that of identifying real-world counterparts to fictional characters as objects of fiction-generated emotions. An important difference, however, is that even where there are no plausible real-world counterparts (as in various types of unrealistic fiction), the art-as-emotional-object view can still hold.

12. Variations on the theme are elaborated in Noël Carroll, *The Philosophy of Horror, or Paradoxes of the Heart* (New York: Routledge, 1990); Lamarque, *Points of View;* Susan L. Feagin, *Reading with Feeling: The Aesthetics of Appreciation* (Ithaca, NY: Cornell University Press, 1996); Edward Gron, "Defending Thought Theory from a Make-Believe Threat," *British Journal of Aesthetics* 36 (1996): 311–12; Eva M. Dadlez, *What's Hecuba*

to Him? Fictional Events and Actual Emotions (University Park: Pennsylvania State University Press, 1997); and Robert J. Yanal, *Paradoxes of Emotion and Fiction* (University Park: Pennsylvania State University Press, 1999).

13. Lamarque, "Fiction," 388.

14. Gron, "Defending Thought Theory," 311–12.

15. Lamarque, "Fiction," 388.

16. For those interested in whether artificial intelligence is possible and, if so, what that might mean for the philosophy of mind, a place to start is Jason Holt, *Blindsight and the Nature of Consciousness* (Peterborough, ON: Broadview Press, 2003), 100–101, 143 n. 4.

17. See Barrie Paskins, "On Being Moved by Anna Karenina and *Anna Karenina*," *Philosophy* 52 (1977): 344–47; and William Charlton, "Feeling for the Fictitious," *British Journal of Aesthetics* 24 (1984): 206–16.

18. Lamarque, "Fiction," 388 (italics in original). Lamarque cites Bijoy Boruah, *Fiction and Emotion: A Study in Aesthetics and the Philosophy of Mind* (Oxford: Clarendon Press, 1988).

Part 3

Brave Newer World: Science Fiction Futurism

THE DIALECTIC OF ENLIGHTENMENT IN *METROPOLIS*

Jerold J. Abrams

> Myth is already enlightenment, and enlightenment reverts to mythology.
>
> —Theodor Adorno and Max Horkheimer,
> *Dialectic of Enlightenment*

Early Science Fiction Cinema

For anyone living in the late twentieth and early twenty-first centuries, science fiction cinema is one of the few art forms that attempt to predict the future of human nature and civilization—a future filled with space travel, nanotechnology, genetic engineering, and widespread surveillance. Ridley Scott's *Blade Runner* (1982) and *Alien* (1979), George Lucas's *Star Wars* sextology (1977–2005), the Wachowski brothers' *The Matrix* (1999), and Stanley Kubrick's *Dr. Strangelove* (1963) and *A Clockwork Orange* (1971) are among the most influential science fiction films. None of them, however, can be properly understood without an initial grasp of Fritz Lang's early science fiction masterpiece, *Metropolis* (1927), based on Thea von Harbou's novel of the same name (von Harbou also cowrote the screenplay).[1] Indeed, contemporary directors borrow quite self-consciously from Lang's early masterpiece: for example, the artificial, gloved black hand of the mad scientist Dr. Rotwang (Rudolf Klein-Rogge) becomes Dr. Strangelove's (Peter Sellers) artificial gloved black hand in *Dr. Strangelove* and then Luke Skywalker's (Mark Hamill) in *Star Wars* (and Anakin Skywalker's [Hayden Christensen] in the second and third episodes).[2] C-3PO is an almost exact copy of the Machine Woman of *Metropolis*. Similarly, Lang's Machine City and underground revolutionary workers become the Machine City and revolutionary workers of *The Matrix* and the machine world of *Dark City* (Alex Proyas, 1998). And the man in the high tower corporation and the Machine Woman, the "false Maria" of *Metropolis,* become Tyrell of the eponymous corporation and the replicants of *Blade Runner.* The list of references and homages goes on and on, and there's no doubt about it: for all great science fiction filmmakers of the twentieth century, *Metropolis* is a cinematic bible.[3]

The Dark City of Metropolis

Set a hundred years into the future (around 2026), *Metropolis* is a silent film (with accompanying musical score) that depicts a dark vision of a dystopian city of the future. The city of Metropolis is modeled on New York as the filmmaker imagined it would appear a century into the future—which today looks like something straight out of *Blade Runner:* skyscrapers everywhere, jets flying at all levels, monorails linking every other floor, and helicopters hovering in the midst of all of it. Joh Fredersen (Alfred Abel) is the designer of Metropolis, a city built for upper-class elites, who live in leisure and pleasure. Among these elites is Joh's son, Freder Frederson (Gustav Frölich), handsome, good natured, and the most privileged young man in all Metropolis. Yet he is also quite naive and foolish. He has no idea about the system in which he lives, except that he enjoys it, spending all his time relaxing and enjoying sex and sports.

Led by Joh, the city of Metropolis runs on the backs of slave laborers, who live in the City of the Workers. There is no middle or lower class, only masters (with a master leader) and slave laborers. We see these laborers walking in rows and descending on an elevator to work underground. As one group of workers descends, another ascends; work and rest. During rest, however, some organize in the city's two-thousand-year-old catacombs, listening to Maria (Brigitte Helm), who is both the leader of an underground movement and the overseer of all the children while the laborers are sleeping and working.

On a field trip, Maria takes the children to a beautiful garden where the elites enjoy themselves. Here Freder's and Maria's eyes meet for the first time (but at a distance). Immediately, Freder is in love, and as she descends, Freder chases her below. But he quickly loses track of Maria and winds up in the great machine hall, standing before the mammoth M-Machine. Again, workers toil, dirty and exhausted. Freder now understands that underneath the sublime magnificence of Metropolis is a dark and evil underworld supporting it. His innocence shattered, Freder is no longer the happy-go-lucky fool. He now knows the torture his father inflicts on his fellow men and women. Immediately, he re-ascends to challenge his father face-to-face:

> Freder: And where are the people, father, whose hands built your city?
> Joh: Where they belong.
> Freder: In the depths? What if one day those in the depths rise up against you?

Freder's words are prophetic. The workers are, indeed, organizing to rise up against Joh. Evidence of the plan has already been found in one of the workers' pockets by the chief foreman of the Heart Machine, Grot (Heinrich George). Grot is loyal to Joh and turns over the plans as soon as he finds them. Joh is grateful, but he cannot decipher them. So he must go see his old friend, the scientist Dr. Rotwang. There are still bad feelings between Joh and Rotwang because both loved a woman named Hel, who chose Joh over Rotwang, and ultimately died in childbirth with Freder. The loss of Hel has destroyed each man in his own way. Joh has become a totalitarian monster, and Rotwang a mad scientist—both from a broken heart. All these years, Rotwang has been toiling away in his sadness and anger, building a new kind of woman to replace Hel, a woman who will never leave him and who will never die. This is *Metropolis*'s most famous character, the Machine Woman, a life-sized female robot. Devilishly proud of his new creation, Rotwang now requires only a persona (a soul and heart, taken from another human being) to complete her. Seeing her in the laboratory, Joh is immeasurably impressed, and Rotwang is happy that he is. The two then retire to Rotwang's study, where Rotwang translates the plan of the workers for Joh. Rotwang tells Joh that they are organizing now in the catacombs, and Rotwang's study just so happens to have a trapdoor leading directly there. They descend together.

Meanwhile, Freder—heartbroken at his father's cruelty toward the workers—has also returned to the underworld, and he offers to trade places (and clothes) with a worker to give him some relief. During a break, Freder finds a copy of the plans in the worker's clothes. Soon another fellow worker notifies Freder that they will be gathering later in the catacombs. So Freder goes and listens to Maria, as do Joh and Rotwang (hidden in an adjacent cave). Maria tells of a coming savior, a great "heart" (a mediator) who will reconcile the "head" (Joh) and the "hand" (the workers). The workers listen attentively but question Maria's strategy of waiting and continuing with their work. They want to revolt against Joh, not wait for a mediator. "We will wait, Maria! But not much longer!" Of course, the mediator will be Freder, who now waits for the crowd to disperse so that he can meet Maria. Immediately, Maria knows the prophecy has been fulfilled. The mediator has come at last. They kiss and agree to meet later in secret.

Now Joh is fully abreast of the threat to his empire and the incredible power Maria wields over the workers. At once, he remembers Rotwang's Machine Woman and his mention that a human soul and heart must be transferred to the Machine Woman. He decides to have Rotwang copy Maria into the Machine Woman (and, once again, take the woman Rotwang loves for his own). The Machine Woman, as a new Maria, will then change her message and sow discord among the workers so that they cannot organize. Of course, Rotwang obeys (or appears to obey), but he too has plans of his own. For what Rotwang sees, and Joh doesn't, looking through a small hole in their adjacent cave, is that the man speaking with Maria is Joh's son, Freder. Here, at last, is Rotwang's chance for revenge on Joh, for which he has been waiting all these many years.

Proceeding as planned (for the moment), Rotwang kidnaps Maria—as Joh orders—thereby intercepting her before she can meet with Freder. Maria cries out to Freder, and Freder wants to save her, but it's no use. Maria is quickly whisked off to Rotwang's laboratory, where she is laid in a tubular chamber in front of the Machine Woman, who sits, icy and motionless, in a chair atop several steps. Rotwang throws a switch, and many glowing hoops scan the women's bodies, copying the soul and heart of Maria into the Machine Woman (which does not appear to harm Maria at all). What emerges from the fusion of the Machine Woman and Maria is now a kind of cyborg woman, that is, the false Maria. Quite the opposite of the pure and innocent true Maria, the false Maria is a dark and sexual demon. In order to show her off, Rotwang takes the false Maria to a men's club. Here she performs an erotic seminude dance before all the ogling eyes of the elite men of Metropolis. Freder is the one elite man who is not there. But, in fact, he too can see the erotic dance, through some faculty of the imagination that is somehow capable of seeing into other scenes in the film—and he's horrified.

Joh, by contrast, is quite delighted with the power of the false Maria, and he believes she is the solution to his problems with the workers. But, again, Rotwang has other plans. He ignores Joh's orders and programs the false Maria to incite the workers to revolt and destroy Metropolis's entire machine underground. Once this destruction has begun, the workers realize too late that they have flooded the city and (possibly) drowned their children. Now, Freder—who has been convalescing from terrible hallucinations—goes in search of Maria and finds her. But the workers, realizing they've been duped, have found her first, and they strap her like a witch to a stake and burn her to death. Freder watches, horrified, but soon he and all the workers realize she is only a robot. With only moments to spare, Freder finds the true Maria, and quickly the two rush to save the children from drowning. Now, with the foundations of Metropolis destroyed, Joh and the workers come together and are reconciled by the mediator, Freder (and the true Maria), to negotiate plans for a new city of the future.

Even eighty years later, audiences are still awed and left dumbstruck at the breathtaking cityscapes, the revolutionary cinematography, and the groundbreaking imagery of the birth of artificial intelligence. And no one can view *Metropolis* without the sure sense that she has just been witness to a sublime philosophical masterpiece. But exactly what Lang's philosophy amounts to is, at least at first blush, far from clear. The question arises frequently: what is the film *Metropolis* trying to say, at its deepest philosophical level?

Dialectic of Enlightenment

The answer to this question lies in another mid-twentieth-century analysis of the rise and fall of modernity, namely, *Dialectic of Enlightenment,* written by Theodor Adorno and Max Horkheimer. This is the master work for all German critical theory (which today includes, most prominently, Karl-Otto Apel and Jürgen Habermas). And it still represents one of the most thoroughgoing critiques of modern technology, instrumental rationality, capitalism, popular culture, fascism, and totalitarianism. On a further (and not unrelated) note, Adorno and Lang were good friends who shared similar views on the demise of the Enlightenment project: they believed that science and technology are not, as is often claimed, forces of liberation from a dark mythological past but rather instruments of a new oppression.

THE THESIS OF THE BOOK

As a preliminary note, Adorno and Horkheimer's text is notoriously—and intentionally—difficult to read and interpret. The ideas are strange and move in incredibly broad sweeps. So what follows is intended not as a master key to their work but only as an interpretation. Adorno and Horkheimer state their central problem in the first lines of their book as follows: "Enlightenment, understood in the widest sense as the advance of thought, has always aimed at liberating human beings from fear and installing them as masters. Yet the wholly enlightened earth is radiant with triumphant calamity."[4] By *calamity* the authors have in mind, and are writing at the height of, Nazi totalitarianism. They want to know why exactly the project of Enlightenment—whose noble aim is the liberation of the human spirit—could lead so quickly, and disturbingly, to fascism. Many today think Nazi totalitarianism was simply a distorted detour, an accident on the way to complete Enlightenment. But Adorno and Horkheimer argue that it was a natural outgrowth of Enlightenment reason. To establish this thesis, they use a distinctly dialectical methodology, though one dramatically contrary to G.W. F. Hegel's dialectical vision. Rather than positing a progression of stages of history toward absolute self-consciousness and freedom, Adorno and Horkheimer see a progression of stages of history toward absolute madness, perfectly embodied in twentieth-century totalitarianism.

FROM HOMERIC MYTHOLOGY TO MODERNITY

Adorno and Horkheimer use two basic concepts: myth and reason, which are in constant tension. Myth is narrative, the stories we tell about our origins, stories about seemingly uncontrollable nature, wild and filled with spirits and gods. Reason, by contrast, controls these dark forces and overcomes the gods. It is calculation and control: "Reason is the organ of calculation, of planning; it is neutral with regard to ends; its element is coordination."[5] And further, "Technology is the essence of this knowledge."[6] The tension between myth and reason first emerges in Homer's *Odyssey*. Odysseus is the original man of reason, who uses technical rationality to manipulate his surroundings despite the wild forces of nature. This is precisely what Adorno and Horkheimer mean by *Enlightenment*—namely, reason's technical overcoming of nature and mythology: "Enlightenment's program was the disenchantment of the world. It wanted to dispel myths, to overthrow fantasy with knowledge."[7] And once begun, Enlightenment reason continues throughout history to subordinate myth in so many progressive stages—a further major stage being Christianity, which provides more organization, more unity at the top, more planning (for example, design throughout the natural world). As Adorno and Horkheimer put it, "The local spirits and demons had to be replaced by heaven and its hierarchy."[8] So the ancient gods are transformed into a singular rational God (a designer god who makes the world rational).

But, of course, Christianity would also come under rational criticism, as yet another myth. In modernity, as Adorno and Horkheimer claim, "all gods and qualities must be destroyed."[9] And "anything which does not conform to the standard of calculability and utility must be viewed with suspicion."[10] So, in place of the Church and heaven and hell, angels and devils, transubstantiation, immaterial souls, damnation and salvation, modernity creates a new culture based in tightly organized states, which are scientific, technical, and highly orderly, perfectly predictable and controllable. Everything is conceived as on a grid.

THE MARQUIS DE SADE'S JULIETTE

To capture the essence of the modern Enlightenment, Adorno and Horkheimer devote an entire chapter of *Dialectic of Enlightenment* ("Juliette or Enlightenment and Morality") to the Marquis de Sade as the quintessential Enlightenment thinker, whom Adorno and Horkheimer conceive as being as important as Immanuel Kant: "More than a century before the emergence of sport, Sade demonstrated empirically what Kant grounded transcendentally: the affinity between knowledge and planning which has set its stamp of inescapable functionality on a bourgeois existence rationalized even in its breathing spaces."[11]

Sade declares that all religion is a myth. Reason is both its destroyer and survivor—as well as the great liberator of the human spirit. These are not uncommon views in modernity. What sets Sade apart, however, is his view of sex. Western sexual ethics have been shaped by Christianity. So, once Christianity is declared a myth, Christianity's sexual ethics are as well. The old rules of heterosexuality, monogamy, love, marriage, etc. are completely gone. Now reason steps in, with no other ends than the rational planning and organization of our most basic animal sexual impulses—hence *Juliette*'s constant and highly organized orgies, which have teams and rules and reasons for every technique. As Adorno and Horkheimer put it, "The precisely coordinated modern sporting squad, in which no member is in doubt over his role and a replacement is held ready for each, has its exact counterpart in the sexual teams of Juliette, in which no moment is unused, no body orifice neglected, no function left inactive."[12] Adorno and Horkheimer's point here, in using Sade, is that every facet of life, even sex, can be seen—at the height of Enlightenment—in terms of absolute planning, absolute control, and absolute rationality.

TOTALITARIANISM

All of this absolute planning, control, and rationality, not unexpectedly, ultimately lead to absolutely controlled political states. And this is why Adorno and Horkheimer claim that "Enlightenment is totalitarian."[13] Now, it's true that Kant argued that "planning" (or hypothetical and instrumental reason) is fine so long as human persons are respected as ends in themselves. But then, of course, that idea (of pure human reasoners) has also been declared a myth. So, just as the Enlightenment used to plan everything *for* human life, now it begins to plan *over* humans, as though they were merely tools in the plan. As Adorno and Horkheimer put it, "Once the movement [of Enlightenment] is able to develop unhampered by external oppression, there is no holding it back. Its own ideas of human rights then fare no better than the older universals."[14] So the individual person is automatically expendable for whichever mythmaker and planning committee happens to be in power.

REASON REVERTS TO MYTHOLOGY

As the possibility of an absolute political state comes onto the horizon, simultaneously reason begins to undergo a kind of identity crisis. It begins to "revert" to mythology, as Adorno and Horkheimer put it. This is perhaps the most mysterious part of their book, that is, how exactly reason becomes myth again. That technological reason can create a violent and totalitarian world is not so difficult to imagine, and many philosophers since have claimed the same. So what exactly do Adorno and Horkheimer mean by *reversion?* Three things, I think.

Reason Declares Itself a Myth: Nietzsche   First off, reason had always defined itself in relation to—and against—mythology. And now, with no mythology left, reason goes searching for a new target, and (finding nothing) eventually it turns upon itself. After all, in attacking myth, reason had always attacked the gods of the day (be they Homeric or Christian). And now the only "god" left is reason itself. So reason turns upon itself with the same aggression and violence it used on the beliefs of the Middle Ages. This is what Adorno and Horkheimer mean when they claim, "Ruthless toward itself, the Enlightenment has eradicated the last remnant of its own self-awareness. Only thought which does violence to itself is hard enough to shatter myths."[15] Adorno and Horkheimer credit Friedrich Nietzsche with the early recognition that reason might—like Zeus, like God—be just one more grand myth, and that reason's essence may be nothing other than "will-to-power."[16] Effectively, reason has killed its very last myth, when it has killed itself—and finally returned to the mythology from which it came. Similar points are made by contemporary philosophers such as Richard Rorty and Alasdair MacIntyre, who claim that human beings are not so much pure reasoners as narrative makers or myth makers.[17]

Return to Mythical View of Nature   Second, reason reverts to myth insofar as our view of nature now appears violent, chaotic, cyclical—as it is in Homer. As Adorno and Horkheimer put it, "Enlightenment has always regarded anthropomorphism, the projection of subjective properties onto nature, as the basis of myth."[18] But once our anthropomorphism is stripped away, what we're left with is a godless, spiritless, disenchanted world, with no forms and no teleological goals. In other words, all that is left is a non-directional, seething, mechanical mass, cyclical and entirely violent—essentially "will-to-power." This is why Adorno and Horkheimer claim that "Enlightenment thereby regresses to the mythology it has never been able to escape. For mythology had reflected in its forms the essence of the existing order—cyclical motion, fate, domination of the world as truth—and had renounced hope."[19] Indeed, the ancient myths now describe reality much better than Enlightenment's original myths of liberation, freedom, and equality. Ancient mythology held up plurality, and now that reason as the will to subordinate plurality has failed, plurality is the case. Mythology held up cyclicality, and now that reason, with its linear conception of history, has failed, and failed precisely because it has returned cyclically to mythology, mythology's cyclicality makes more sense.

Desublimation of Ancient Mythology   As a third form of return, we actually return to the content of the ancient myths (in a way). On this point, there is a certain Freudian dimension of the dialectic of Enlightenment, particularly in terms of a dialectic of sublimation (or repression of drives) and desublimation (release of those repressed drives). The collective unconscious is filled with images of mythology: heroes, gods, monsters, demons, violent nature, magic, and many ancient symbols. And these are repressed and held at bay by reason. Yet, once reason declares itself a myth (in the hands of Nietzsche), and now that reality more closely resembles that of the ancient mythical world, all of those ancient symbols and myths of heroes and demons, all those violent, pluralistic, wild, erotic impulses come bubbling back up to the surface.

Of course, this is not to say that we're returning to Homer's world exactly. Rather, what happens is that ancient mythology is reborn within the totalitarian nation-state. Total organization and mythology fuse into a new and violent world of magic and force. Indeed, on Adorno and Horkheimer's view, it would be no accident that Adolf Hitler would draw upon so many ancient symbols and elements of Nordic mythology to create a new myth of the pure Aryan seed laid in the earth to dominate all the other ape-men, at the very height of Enlightenment reason, when reason—far from achieving its supposed end of absolute consciousness—would begin to collapse into the absolute madness of the Nazi regime.

Metropolis and *Dialectic of Enlightenment*

Lang's *Metropolis* captures perfectly this philosophical vision of *Dialectic of Enlightenment:* out of mythology, into totalitarianism, and back into mythology.

MYTHOLOGY OF THE DARK AGES

Mythology plays an essential role throughout *Metropolis.* For example, recall the two-thousand-year-old catacombs. The symbolism here is fourfold. First, this world is underground, which symbolizes its pastness—like a "buried past," with thousands of skulls and bones from generations of believers and workers long gone. Second, being two thousand years old, it has an obvious connection to the birth of Christianity. And there are, moreover, multitudes of crosses on the rock altar. Third, being underground and almost entirely in the dark, it has a connection to the Dark Ages of medieval Christian thought. Fourth, Maria is also Mary as Madonna. She is pure, dressed in white, clean—and we know she is both (sexually) a virgin and (symbolically) a mother; remember, she takes care of all the workers' children while they are working and sleeping. And her prophecies, similarly, are also drawn from Christianity. She assures the oppressed workers that a savior will come to help them, a mediator who will at last bring peace and balance between the lower and the upper worlds.

MODERNITY

From this dark, underground, mythical past, however, there arises a great Enlightenment birth of reason. Here is the new upper-world city, beaming with light, liberated—not religious, and entirely scientific and modern. As we move from lower to upper, from Middle Ages to Enlightenment, the text scrolls upward and appears in pyramid form—like this:

As

deep as

lay the workers'

city below the earth,

so high above it towered

the complex named the "Club

of the Sons," with its lecture halls

and libraries, its theatres and stadiums.[20]

The scenes that follow might as well have been taken right out of *Dialectic of Enlightenment:* perfectly ordered rituals of sexual behavior, sporting events, science, modern architecture, and high technology. Everything is organized, systematized, and controlled, and diversity is subordinated to the structure of the whole.

On a further silent frame, we also read Metropolis's city motto: "GREAT IS THE WORLD AND ITS CREATOR AND GREAT IS MAN." "Creator" here, however, should not be taken to mean God, but rather Joh, "creator" of all Metropolis—and note well, there simply is no "world" outside the world of Metropolis. Indeed, there is no mention of God among the Metropolis elites, and no mention of religion at all.

METROPOLIS AS TECHNOTOTALITARIAN

As creator, designer, and ruler of Metropolis's technocity, Joh is also dictator. Metropolis is no democracy but a fascist totalitarian state where absolute order is the rule, where the elite enjoy power and privilege and the workers are slaves. These workers are also entirely undifferentiated: they all look alike, they all wear the same clothes, they all walk in the same mechanical way, filing in and out of the Machine City in lines. They appear just as robotic as the levers and pulleys and clocks and pumps they operate. This work, moreover, never stops. Two shifts alternate; as one goes into the Machine City, the other goes out. And as the film historian Enno Patalas points out, those coming from work walk half as fast as those going in (who are already slow from exhaustion). There is no leisure at all. That is reserved only for the wealthy elite, who enjoy free time and exercise in the sports stadium (in a sense). And here I should note that my interpretation diverges from Patalas's commentary: "The sports stadium, the contrast is stark between its openness under sweeping skies, and the cramped City of the Workers—just as stark as the contrast between the liberated and carefree movements of the youths, dressed in white, and the dull lethargy of the darkly clothed workers—and the self-determined horizontal movement versus the downward ride of the workers in the lift."[21] I agree with part of what Patalas is saying here: clearly Metropolis conveys this distinction between the workers and the elite, in general. But, on closer inspection, notice that the contrast betrays as much similarity as difference. In cutting directly from workers to athletes, clearly Lang is conveying their parallels. For the athletes perform physical activity (like the workers), wear identical uniforms (just like the workers), move in lines in their race (just like the workers), follow rules, and do everything in perfect time and order (exactly like the workers). Notice, too, the fascist architecture, with almost no detail (and, indeed, no seating at all)—the lines sweep upward, towering over the athletes, blinding them from the outside world. In fact, this scene of the sports stadium is much in line with the analysis of modern sports and the Marquis de Sade's orgy teams as Adorno and Horkheimer describe them: the "precisely coordinated modern sporting squad, in which no member is in doubt over his role and a replacement is held ready for each."[22]

All of Metropolis operates on this vision. Nothing is wasted. Every moment is accounted for, even in leisure. Even the sexual rites in the Eternal Gardens are strict and organized, as young women are herded about by a kind of headmaster (Patalas calls him a ringmaster) of the garden. Each is turned about for study and selection; one, in particular, turns robotically to show her front, which is fully clothed, while her back is almost completely bare. "Which of you ladies shall today have the honor of entertaining [having sex with] Master Freder, Joh Fredersen's son?" Indeed, Metropolis's fundamental aim is to dissolve humanity into robotics, until the human spirit is gradually lost. And this holds not just for the workers but for everyone in Metropolis (contrary to Patalas's otherwise excellent interpretation).

THE FUSION OF HUMAN AND MACHINE

This fusion of human and machine, moreover, is not restricted to habituation but is ontological as well. For example, Rotwang has sacrificed his own hand to the Machine Woman. He now has a robotic hand (which is always covered in a black glove), made from the material of the Machine Woman. Rotwang is part Machine Woman, and the Machine Woman is part Rotwang. Here is the birth of the cyborg in cinema, a form that will be repeated often but that also has its ancestors. Certainly, Mary Shelley's *Frankenstein* is one, but a more approximate ancestor is the Tin Woodsman from L. Frank Baum's 1900 book *The Wonderful Wizard of Oz.* To be sure, the Machine Woman looks very much like the Tin Woodsman (at least as he appears in the film *The Wizard of Oz*). Equally important, both the Tin Woodsman and the Machine Woman are missing something that would otherwise make them human. The Tin Woodsman is missing a heart. And the Machine Woman, in the Maria-Machine fusion, is shown to have a glowing beating heart, suggesting this is part of what must be transferred to the Machine Woman. One may also note strong similarities between the futuristic cities of Oz and Metropolis and between the Wizard and Joh.

George Lucas, the creator of the *Star Wars* series, clearly loved this theme, and he paid homage to both works, *The Wizard of Oz* and *Metropolis,* in *Star Wars.* C-3PO looks exactly like Lang's Machine Woman and yells out, just like the Tin Woodsman, "Over here"—because he is all stiffened up. He requires an oil bath, which he receives in Luke Skywalker's workshop, just as Dorothy oils the joints of the Tin Woodsman, and just as Rotwang "activates" the stiff and almost lifeless Machine Woman.

From *The Wizard of Oz* to *Metropolis* to *Star Wars,* the machine-human is a science fiction staple and is easily one of its most important icons. It is important for two reasons. First, as we have seen, it serves as a template for all further machine-human fusions in science fiction, including C-3PO, Darth Vader, the replicants of *Blade Runner,* the Borg and Data in *Star Trek: Generations* (David Carson, 1994), the Terminator, HAL in *2001: A Space Odyssey,* the Mecha in *A.I.: Artificial Intelligence* (Steven Spielberg, 2001), Robocop in *Robocop* (Paul Verhoeven, 1987), and even the fembots, parodies of the Machine Woman, in *Austin Powers: International Man of Mystery* (Jay Roach, 1997). Second, the Machine Woman also provides a template for the future of real-life technology. Today, many scientists are taking the idea of a Machine Woman quite seriously. In the literature on artificial intelligence, the process of transferring a human mind, that is, copying its exact atomic structure, into a robotic receptacle is called *uploading.* Obviously, uploading is currently not technically possible, but many believe it will be possible not long after the time Lang predicts. The contemporary analogue of the Machine Woman is a design created by a group of scientists, philosophers, and AI designers known as the Extropians. Their design is called the Primo Posthuman (and, perhaps not unintentionally, it looks eerily like the Machine Woman from *Metropolis*).[23]

REASON TURNS UPON ITSELF

Having reached such a point of perfect rationality, absolute control, and efficiency, however, the system of Metropolis steadily begins to turn upon itself. The point is evident partly in Rotwang, who embodies instrumental rationality and turns upon Joh. But it is perhaps most evident in Freder, who turns upon his father. Freder has, until recently, idolized his father as a god of pure reason, just as reason idolizes itself in modernity. But, now, Freder has discovered the dark secret kept fast from the elite. Metropolis is not great, and it continues to live a lie. Most of the people of Metropolis do not know that the workers lead such horrible lives. Joh enforces this secret because such knowledge would negatively affect their self-adulation and supposed perfect harmony. But now Freder sees that a new idol has simply replaced the old one, and the new one is no better. In place of the God of the Middle Ages, now the new totalitarian god of Metropolis is equally empty. Freder has hollowed out this new god, just as Joh hollowed the last god (and created Metropolis). A violent god still stands high in his castle, lording over all the little people, much as the Holy Roman emperors stood over all the feudal masses and the Egyptian pharaohs stood over the Jews.

METROPOLIS RETURNS TO MYTHOLOGY

But of all the elements of *Metropolis* that resemble Adorno and Horkheimer's vision, perhaps the most striking is the fusion of technototalitarianism and ancient mythology. At once, we see an advance into the future total technological order and a corresponding reversion to ancient mythology, perhaps nowhere more evident than in the character of Joh, the new god of Metropolis. As Patalas points out, Joh's name is spelled with an *h*, intended as a variation on the Old Testament name Jehovah. And Joh is also a hard and violent god, as well as the creator and designer of Metropolis. He is even called the "brain of Metropolis," always watching his clocks, a god of time and control. Joh's deceased wife is also a god or, rather, goddess—Hel, who in Norse mythology is the Queen of Hel (they share the same name), the Norse underworld. Note, too, the very name Metropolis means "mother city" (as in Hel's city).

Joh also resides high above all Metropolis, in the highest building in the city, the New Tower of Babel. In Genesis 11, the people built the Tower of Babel to reach the sky and make humanity like God. But God was angered by it and destroyed the tower and punished the makers by fragmenting the universal language into all the languages of Earth. Here, however, it is the creation of the tower that has divided the people (workers from elites). And at the height of Metropolis's reign, this city will also be destroyed—as Maria rightly prophesies in the catacombs below. She tells the ancient myth of the Tower of Babel. The implication could hardly be missed, that the New Tower of Babel is the same thing all over again, merely reborn within Joh's new totalitarian Enlightenment.

The fall of Babel begins with Freder's descent from the Eternal Gardens, which represent a new Garden of Eden. These gardens are beautiful and organic, created by the god Joh especially for his son, Freder (like an Adam). Here Freder enjoys himself among trees and fountains, with many beautiful maidens prancing around him. He can sexually indulge himself to his heart's desire, with no idea of what is happening behind the scenes. But as Freder learns of good and evil, he leaves the gardens and descends into the underworld, where he sees the M-Machine. Patalas notes that *M* stands for "mother" and perhaps, too, for "Moloch, the God of the Ammonites, to whom the Israelites also sacrificed children, to the chagrin of Moses and his God." Staring in horror at the M-Machine, Freder begins to hallucinate and to create a kind of film within his mind. The M-Machine becomes a sublime fusion of futuristic technology and ancient mythology. All manner of cranks and pulleys support a giant, lionlike monster, with its furious eyes and gaping hungry mouth (about twenty feet tall), with giant front arms and paws. A staircase leads between the M-Monster's resting paws, and Egyptian-dressed guards herd workers up the steps to be thrown into the mouth and burned alive in the flames.

A similar fusion of technology and ancient mythology occurs with Dr. Rotwang and the false Maria. Rotwang's laboratory, in particular, is filled with super-high technology and numerous mythological symbols and medieval gadgets. Everything here is either ancient or futuristic—even his home is ancient in style (though built in the midst of Metropolis). Nothing here in Rotwang's world is strictly modern. But perhaps the most obvious fusion of the distant past and the distant future is seen in Lang's shot of the step altar upon which the Machine Woman sits underneath a giant, inverted pentagram (an ancient mythical symbol of evil). The Machine Woman is to be part ancient, mythological evil and part futuristic technology. And the machine-mythology synthesis is intensified once the uploading of Maria is complete. For now Rotwang can take the false Maria to the Club of the Sons, where she performs her erotic dance. Freder, as we noted, can see this dance (through some faculty of mind), even though he is far away from the club. At first he sees what the other men see—the erotic dance. But then, as in the case of the M-Machine, through his cinematic imagination, Freder sees the dance as a return to ancient mythology. The false Maria becomes the whore of Babylon (from Revelations 17), riding upon a great seven-headed beast. Freder has several of these hallucinations, including visions of the seven deadly sins and of the grim reaper coming to kill everyone.

And here at the penultimate moment of *Metropolis*, when totalitarianism and mythology are completely fused, everyone appears absolutely mad—just as Adorno and Horkheimer claim is the ultimate end of Enlightenment. Freder, of course, is sick and convalescing from wild bouts of hallucinations. Joh, the cruel dictator, is mad with power. Rotwang (in the tradition of Dr. Frankenstein) is an absolutely mad scientist, determined to destroy his world. The false Maria appears insane on stage, as Lang's camera portrays a hundred floating, whirling eyeballs of the sons, mad with lust. And the workers, no longer rational people but zombies, are stirred to a mad frenzy by the false Maria to destroy Metropolis—failing to realize that such an act might ultimately kill all of their children.

Metropolis Eighty Years Later

Eighty years after its release, *Metropolis* remains a remarkably sublime work of art, absolutely breathtaking in its scope and grandeur—and, not least of all, its eerie prophetic vision. Lang presented perfectly, in some of the greatest moving images ever put to screen, the entire philosophical movement of Adorno and Horkheimer's *Dialectic of Enlightenment*—anticipating them by almost two decades. Living in Hitler's Germany, Adorno and Horkheimer watched the Enlightenment become totalitarian, revert to mythology, and descend into insanity. As witnesses to the rise of Hitler's fascist totalitarianism, they had the philosophical benefit of watching it happen. But Lang envisioned the same mad totalitarian-mythology fusion long before Hitler came to power. Few films—indeed, few works of art—can claim such clarity of philosophical vision. Most great systems of thought must be formulated after history has already happened, as Hegel once claimed. But somehow Lang understood precisely how the Enlightenment, as essentially a philosophy of technological reason, would lead directly (and seemingly unwaveringly) not into absolute mind, as Hegel had claimed, but instead into absolute madness.

Notes

I would like to thank Steven M. Sanders for reading and commenting on earlier drafts of this essay and Elizabeth F. Cooke for conversations on both *Metropolis* and *Dialectic of Enlightenment*. The epigraph to this essay is taken from Theodor Adorno and Max Horkheimer, *Dialectic of Enlightenment: Philosophical Fragments*, trans. Edmund Jephcott, ed. Gunzelin Schmid Noerr (Stanford: Stanford University Press, 2002), xviii.

   1. The original film of *Metropolis* is no longer intact. Many scenes are missing, and the work as a whole has been heavily edited. As the film historian Enno Patalas has pointed out, "More than a quarter of this film must be regarded as irretrievably lost. Few films have been so systematically changed, mutilated, or corrupted as this one. Shots and titles have been omitted and changed. However, of no other such mistreated film do we know so well what the film originally looked like." Enno Patalas, commentary, *Metropolis*, DVD (1927; restored ed., New York: Kino International, 2003).

   2. The theme of the robotic hand and arm is repeated regularly; see, for example, *The Terminator* (James Cameron, 1984) and *Terminator 2* (James Cameron, 1991).

   3. Roger Ebert, in a review of *Metropolis*, also draws out these points when he writes, "The movie has a plot that defies common sense, but its very discontinuity is a strength. It makes 'Metropolis' hallucinatory—a nightmare without the reassurance of a steadying story line. Few films have ever been more visually exhilarating." He continues, "Generally considered the first great science-fiction film, 'Metropolis' (1926) fixed for the rest of the century the image of a futuristic city as a hell of scientific progress and human despair. From this film, in various ways, descended not only 'Dark City' but 'Blade Runner,' 'The Fifth Element,' 'Alphaville,' 'Escape from L.A.,' 'Gattaca,' and Batman's Gotham City. The laboratory of its evil genius, Rotwang, created the visual look of mad scientists for decades to come, especially after it was mirrored in 'Bride of Frankenstein' (1935). And the device of the 'false Maria,' the robot who looks like a human being, inspired the 'Replicants' of 'Blade Runner.' Even Rotwang's artificial hand was given homage in 'Dr. Strangelove.'" Roger Ebert, review of *Metropolis*, http://www.ebertfest.com/four/metropolis_silent_rev.htm.

   4. Adorno and Horkheimer, *Dialectic of Enlightenment*, 1.

   5. Ibid., 69.

   6. Ibid., 2.

   7. Ibid., 1.

   8. Ibid., 5.

   9. Ibid.

   10. Ibid., 3.

   11. Ibid., 69. Immanuel Kant authored *Critique of Pure Reason* (1781), widely acknowledged as one of the most important works of modern philosophy. Here Kant develops his famous "transcendental philosophy." Kant also applied his transcendental method in his moral writings and attempted to transcendentally deduce the moral law

from pure practical reason. But that, as is well known, and as Adorno and Horkheimer point out, is hardly without its problems. What remains, then, after Kant's categorical imperative is shown to be ungrounded is one's own expertise in planning and forming hypothetical imperatives. And this is what Adorno and Horkheimer have in mind in their analysis of Kant as a philosopher of total planning—hence the analogy to Sade.

12. Adorno and Horkheimer, *Dialectic of Enlightenment,* 69.

13. Ibid., 4.

14. Ibid., 3.

15. Ibid., 2.

16. Ibid., xviii.

17. Richard Rorty, *Contingency, Irony, and Solidarity* (Cambridge: Cambridge University Press, 1995); Alasdair MacIntyre, *After Virtue* (Notre Dame, IN: University of Notre Dame Press, 1981). MacIntyre writes, "A central thesis then begins to emerge: man is in his actions and practice, as well as in his fictions, essentially a story-telling animal. He is not essentially but becomes through his history, a teller of stories that aspire to truth." MacIntyre, *After Virtue,* 201.

18. Adorno and Horkheimer, *Dialectic of Enlightenment,* 4.

19. Ibid., 20. This is also why so many postmodern philosophers today emphasize that culture has a kind of nonlinear feel to it, a kind of mechanical reproduction of forms, a repetitive emptiness. Adorno and Horkheimer make this point on a more theoretical level when they write, "The more completely the machinery of thought subjugates existence, the more blindly it is satisfied with reproducing it." Ibid. For example, even at the very minor level of everyday entertainment, "every film is a preview of the next, which promises yet again to unite the same heroic couple under the same exotic sun: anyone arriving late cannot tell whether he is watching the trailer or the real thing." Ibid., 132.

20. Patalas quotes the filmmaker Luis Buñuel: "Even the titles, how they rise and fall, blend with the movement as a whole, become pictures themselves." Patalas, commentary.

21. Ibid.

22. Adorno and Horkheimer, *Dialectic of Enlightenment,* 69.

23. See http://www.natasha.cc/primo3m+diagram.htm.

Imagining the Future, Contemplating the Past

The Screen Versions of *1984*

R. Barton Palmer

> We can talk to the past as we can talk to the future—the time that is dead
> and the time that has not yet been born. Both acts are absurd, but the
> absurdity is necessary to freedom.
>
> —Anthony Burgess, *1985*

A Startling Alterity

A defining feature of science fiction is that such works of imaginative realism (a potent stylistic brew of perhaps irreconcilable elements) speculate
about some future age or alternative, extraterrestrial world. That imagined
place and time is characterized essentially by "advancements" in science
that plausibly explore the consequences of what is now known and actively
researched (in such areas as artificial intelligence, genetic manipulation,
space travel, pharmacology, and so forth). The difference between the reader's
implied present and the postulated alternative results from the technological
manipulation of the natural environment and human experience that such
acquired knowledge makes possible.

This difference encourages musings that constitute a principal form of
readerly pleasure, with science fiction characteristically focusing more on
"representation" broadly considered than either narrative or character. As
many commentators have pointed out, in fact, this concern with representation also often figures dramatically, through the opposition of principal
characters whose conflict is not one of motive or disposition but ideology

or worldview more generally. Having staged an unfamiliar reality, science fiction customarily provides it with depth through an anatomizing of its sustaining ideas and principles. Such exposition unfolds agonistically, as a contentious dialogue in the tradition of Menippean satire. Dystopian fiction, that is, science fiction that deals with a nightmarish future, particularly emphasizes this aspect of the Menippean form, offering, as Northrop Frye suggests, "a serious vision of society as a single intellectual pattern." Such fiction, Frye observes, deals "less with people than with mental attitudes . . . with abstract ideas and theories."[1]

Dystopian fiction, of course, is conventionally realist to the extent that such texts confect worlds recognizably like our own, to the extent at least that they are inhabited by human beings similar in essential ways to us. Yet these imagined settings are also marked by an often startling alterity, a "difference" that is the textual figuration of fear or desire or perhaps, more commonly, a complex mixture of the two. Dystopian fiction presents us with futures that conform to our deepest terrors—and wishes. An important point is that we owe to the Enlightenment concept of progress, confirmed by much of the experience of the twentieth century, an acceptance of this kind of future. Enlightenment optimism about the inevitable malleability of nature and human nature provokes the expectation of a succession of states of affairs strikingly, substantially, and unpredictably distinct from the present.

Such futures are not the products of some process defined by the glacial tempo of the *longue durée*, the kind of slow-moving historical change that led, for example, to the advent of capitalism. In staging and then dissecting such speculative futures, dystopian fiction by its very nature engages critically with the cultural project of the Enlightenment, so dependent on the triumphalist rationalism that underpins scientific thought and inquiry. Such fiction is thus essentially an anti-Enlightenment form, the "abstract ideas and theories" it takes up and contests deriving from the application of reason to the shaping of human experience and the redesign, for human advantage, of the natural world.

I have rehearsed these essential points about dystopian fiction because arguably the best known work in that genre, George Orwell's classic novel *1984* (1949), whose two screen adaptations (Michael Anderson, 1956; Michael Radford, 1984) are the main focus of this essay, has often been read chiefly as a complex, moving meditation on twentieth-century totalitarianism, to the neglect of the critique it offers of Enlightenment thought more generally and especially the notion of inevitable progress (or, less optimistically, sudden and disastrous change arising from the continuing reification and rationalization of human relationships).[2] Orwell's critique of Enlightenment thought was substantially influenced by his reading of anti-Enlightenment thinker James Burnham. And yet, so impressed by *1984*'s seeming political topicality, its connection to the disillusionment of the early Cold War era, many critics have been inclined to see Orwell's project less as a critical engagement with Burnham's most famous text, *The Managerial Revolution* (1941), than as a disenchanted socialist's extended rant against the heartless excesses of Stalinist central planning and deployment of surveillance and terror as mechanisms of social control.[3]

For Burnham, the failure of capitalism would come not through a socialist revolution but because of the ongoing separation of ownership (increasingly diffused throughout a body of stockholders) from overall control over the means of production. Such control was in the course of the twentieth century becoming increasingly exercised by a class of managers who, in Burnham's view, would quickly be the only ones in society capable of governing an increasingly technologically sophisticated and organizationally complex world. In short, Fordism, not Leninism, would prove triumphant, and Frederick Taylor, one of the fathers of modern management theory, would be more influential in shaping the future development of society than Karl Marx. Interestingly, a similar perspective, anticipating Burnham's critique and Orwell's response, underlies the future world limned by Aldous Huxley in *Brave New World* (1932). Burnham's views were widely influential in the period, even domesticated and popularized by such tamer apocalyptic visions as William Whyte's *The Organization Man* (1956), a polite rant against the homogenizing effect of modern corporate employment on lower-level managers, and C. Wright Mills's *The Power Elite* (1956), which draws a scathing portrait of their upper-level counterparts in what Dwight D. Eisenhower identified as the "military-industrial complex."

However much he was proved correct by some of the developments in the immediate Cold War era (including the division of the world into superpowers and their spheres of influence), Burnham, Orwell declared, was self-deceptively preoccupied with "monsters and cataclysms," and this is the reason he was inclined "to overrate that part played in human affairs by sheer force."[4] As William Steinhoff suggests, what Orwell did in *1984* was to "project the logical implications of Burnham's arguments into the future" in order "to refute them," though the precise nature of this refutation is perhaps disputable since the novel ends rather dismally with the complete "re-education" of its two erstwhile rebels and thus a reaffirmation of the total power residing in the system.[5] In any event, Orwell most thoroughly anatomizes Burnham's vision in *The Theory and Practice of Oligarchical Collectivism*, the condemned "book" by "Emmanuel Goldstein." Goldstein is the imagined Trotskyite counterrevolutionary who, in the spirit of Marxist analysis, demystifies the world of *1984*, exposing its nationalist and collectivist ideology as a sham. Goldstein and his book, of course, ironically enough, are part of the same sham, constituting as necessary fictions a confected opposition to the current regime, which has collectively contrived at their creation. The book is the lie that is also truth, the truth that is also a lie. In *1984*, what Orwell regarded as the impossibilities of Burnham's predictions could be brought frighteningly to life, especially the evolution of a world in which a managerial class, misrecognized as a "party," has come to control the means of production with the simple, but disguised, end of perpetuating its own exercise of power.

Self-Conscious Nostalgia

Like the beast epic genre he resurrects for *Animal Farm* (1945), which allows him to decontextualize and, to a degree, render more abstract his political argument, the dystopian form Orwell employs in *1984* permits representation to escape the iron laws of reference and historical specificity. Such futurism and abstraction lead Orwell to substitute possibility for plausibility as the basis for a new form of "reality" effect. Made the basis of shared visions, literary and cinematic, such imaginings or projections may prove problematic, however, or so some have thought. Paul Ricoeur, for one, holds the view that the utopian/dystopian impulse in literature and film is ultimately escapist (for him a term of abuse) and rhetorically deceptive, concealing as it does "under its traits of futurism the nostalgia for some paradise lost." In Ricoeur's view, "Escapism is the eclipse of praxis, the denial of the logic of action which inevitably ties undesirable evils to preferred means."[6] This is true enough, I suppose, if we hold a limited view of praxis as what we might do right now to set right a social evil, an intention eclipsed, so to speak, by our imagining of some future moment in which it has either worsened or disappeared. Lewis Mumford, in contrast, offers a more positive cultural explanation for such imaginings, rethinking them as, in effect, a form of praxis: "Every community possesses, in addition to its going institutions, a reservoir of potentialities, partly rooted in its past, still alive though hidden, and partly budding forth from new crossings and mutations, which open the way to further development."[7] There is, in other words, Mumford suggests, a connection between our dreams of the future and what we choose to do in the here and now.

Ricoeur, I think, misses a further central point, one that will be explored at some length in this essay. Imagined worlds hold an immense usefulness for a symptomatic analysis of the present. For if futurist fiction avoids any representation of the contemporary world, it does not follow that it also disengages from its concerns. Escapism, after all, constitutes a double movement that is both away (from the present) and toward (the postulated future). As an element of a historically determined *mentalité*, nostalgia partakes of the moment with which this kind of text engages. It is in fact nothing less than the particular readerly state of mind that forms the basis of its imaginings. It is also the desire to which the resulting fiction appeals. At different times, we imagine the future in different ways, and these divergent visions correspond to our always already contextualized hopes and fears. One reason that the alterity of dystopian fiction is startling, then, is that in such plausible confections we recognize an aspect of the life we are now living, namely what at the moment we both desire and fear. If literary texts depend on the process of "making strange," then dystopian fiction constitutes something like a limit case of such a technique, deploying as it does another world to stand in for our own, beyond which any further enstrangement hardly seems possible.

In the case of *1984*, composed in the aftermath of a partial, ironic collapse of authoritarianism (Hitler and Mussolini defeated to make the world safe for Stalin and Soviet-style communism), the reader is invited to fear the thoroughgoing totalizing and rationalizing of human relations in the not too distant future (postulated as less than four decades distant from the historical moment of its first readers). Anthony Burgess emphasizes the novelist's address to his own era, his defamiliarizing evocation of the world that he shared with his first readers: "Orwell really wasn't forecasting the future. Novels are made out of sense-data, and it's the sensuous impact of the novel that counts to me."[8] True enough, and we are thereby reminded (a point to which we later return) that the novel conjures up a future that is enthralled to its own idea of a future. Orwell thus imagines a world in which the Party has proclaimed that the richness and flexibility of the English language must be reduced to pure, limited utilitarianism in order to eliminate the possibility of "deviant" thought. But that world does not yet exist, and Winston Smith is able to translate his discontent into language that opposes the language of the state; his diary is the concrete evidence of such a communicative possibility. Similarly, Orwell imagines a world in which the Party has determined that the sexual urge, founded on the experience of the orgasm, must be eliminated so that romantic and familial connections may be rendered unlikely if not impossible. The aim is to ensure that no institution mediates between the state and the individual, as procreation, it is projected, will be efficiently and totally rationalized by the practice of artificial insemination. But such a state of transformed nature lies only in the future contemplated by those who inhabit Orwell's future. Winston and Julia, brought together in the novel's present by the possibility of orgasm, do deeply fall in love, their romance constituting a sign that their historical moment is incomplete, not yet delivered to itself.

Orwell also invites his first readers to indulge their desires for, first, a life delivered from absolute subservience to an omnipotent bureaucracy by a decision to turn solipsistically toward the pleasures of one's own thoughts (including the writing of a diary); second, an engagement with the forces of opposition, founded on a scientific analysis of the deceptive workings of the state; and, finally, a thoroughgoing rejection of Victorian moral strictures for a free, self-justifying indulgence in sexual pleasure (a "wish" that looks backward to D. H. Lawrence but also forward to Erica Jong). Living in a postwar Britain still suffering from food shortages and rationing, Orwell's first readers were even invited to contemplate the sudden, antisocial acquisition of luxuries calculated to appeal to the national taste: sugar, tea, coffee, and, most wondrous of all, real chocolate. As J. P. Telotte notes, following Mumford, the projection of desire into an unrealized future constitutes not only a flight from reality but an expression of "dissatisfaction with the world as it is currently constituted and a belief in the possibility of change."[9]

If futurist fiction thus depends on a potent dialectic, attracting viewers by postulating "a fully imagined and convincing world" yet working to "displace us from the real, even toy with our convention-bound sense of reality," it does so in order to play complexly with notions of fantasy and engagement, its rhetoric one of projection and (in the manner of allegory) displacement.[10] In its antiestablishmentarianism and wishful thinking, its general disillusionment with socialism and government, *1984* shares much with such postwar British fantasies as the Ealing comedies (chief among them *Whisky Galore* and *Passport to Pimlico,* both 1949) and early "angry young man" fiction, principally Kingsley Amis's *Lucky Jim* (1954). As Anthony Burgess has pointed out, speaking of *1984*'s engagement with its own time,

"The cheating of the senses with shoddy food, drink and tobacco . . . it was all there for fictional transference. It was a bad time for the body."[11]

I repeat. The imaginative realism to be found in dystopian fiction is a very unstable, perhaps contradictory brew, caught between "times" and representational rhetorics, much like the novel's narrator, who does not know whether to address his book to the past (which can scarcely care about his predicament) or to the future (when the revolutionary energies of Ingsoc will have either triumphed or dissipated, making Winston's extended appeal either irrelevant or pointless). Caught between times, such texts are also only problematically realist. Most futurist fictions neither foster nor depend upon illusionism (the encouragement offered reader and viewer to believe in the textual experience as fundamentally like the "real"). Instead, as Telotte points out, "a major effect of their stylizations is to reinscribe the fundamental question of reality, the sense of how our notion of the real . . . is always implicated in these 'nostalgic' projections."[12] Science fiction of the utopian/dystopian variety, in short, is metafictional, that is, making readers conscious we are reading a text and forcing us to "think" to a lesser or greater degree the fictional and the ideological as cultural categories, as we confront in imagined futures our desire (for the impossible return of a bygone era) and fear (that progress is a destructive illusion).

This idea of a self-conscious nostalgia leads Telotte to consider futurist fiction as "protopostmodern." What prevents this genre from a full engagement with postmodernism, a movement that denies either the fact of or any interest in historical change, is "its postulating of a possible trajectory for human history." And yet, "in its assault on the status quo and the manner in which it couches that assault," futurist fiction "inevitably implicates the question of the real in a postmodern—as well as fantastic—manner."[13] Postmodernism has slipped quotation marks around the concept *history*, challenging the utility of all master narratives of human experience, whereas Orwell's novel, like other examples of its subgenre, issues a warning about the way in which history, following the logic of the Enlightenment, might well unfold. But this is again a very unstable brew, as exemplified by the film versions of *1984*, which extend the reach of the novel to audiences of eras separated by three decades of historical change.

Produced less than a decade after the novel's initial publication, Michael Anderson's film version of the novel still offers Orwell's strong sense of "a possible trajectory of human history." This *1984* is a film that speaks to the future, as the novel does, engaging real fears and issuing something in the nature of a warning. But in Michael Radford's otherwise celebratory and reverential cinematic adaptation, Orwell's prophetic vision has been fully accommodated to postmodernism's preferred form of nostalgia: a desire projected from that moment when ideological conflict and even history itself appear to have ended back into a past when ideas (such as the communal good and individual freedom) still seemed to matter and significance itself, so the dream goes, was still possible. In the realm of the postmodern, Terry Eagleton laments, it is as though "all the high drama, all the self-risking and extravagant expenditure which might have belonged to our moral and political life together in more propitious historical conditions, had now been thrust back into the contemplative theatre of reading."[14] Radford's version of *1984* reconstitutes its fiction as just such a "theatre of reading."

Talking to the Future: *1984* (1956)

Orwell's novel, to be sure, can be read metafictionally because it foregrounds its constructedness in a number of ways, especially through its lack of interest in presenting plausible characters (From what does Winston Smith's rebelliousness stem? Why does Julia choose him as her next lover?); its failure to offer an explanation for the critical events that lead from the end of World War II to the establishment of Ingsoc; its silence about the organization of Oceania as a global whole; and its reluctance to provide any technological explanation for the televisual surveillance system, including two-way communication, that is such a frightening aspect of social control in Orwell's London, now renamed Airstrip One.

These features of the novel, however, have customarily been read as aesthetic flaws, not as deliberate marks of artificiality or constructedness. Typical is Harold Bloom's view that *1984* is a "good bad book," saved only by its "relevance" (its connection to twentieth-century politics) and its rhetoric, suiting the moment of the Cold War through the persuasive, if vague, warning it issues about the future, namely that "the dreadful is still about to happen."[15] Michael Walzer is not unusual in thinking about *1984* less as a novel and more as a philosophical tract that anticipates Hannah Arendt's *Origins of Totalitarianism* (1951) and Czesław Miłosz's *Captive Mind* (1953), rather than following that utopian fictional trend that includes such texts as Edward Bellamy's *Looking Backward* (1881) and H. G. Wells's *The Time Machine* (1895).[16]

By the early 1960s, so John Rodden reports, Orwell's novel had capitalized on the earlier success of *Animal Farm,* which had become a widely assigned high school text, and had begun to undergo the same kind of canonization (despite what many saw as the novel's objectionably amoral romance).[17] At that time, the book was beginning to be read within the context of other "topical fiction" dealing with the Cold War, including such dystopian novels as *The Bedford Incident* (Mark Rascovich, 1963) and *Fail-Safe* (Eugene Burdick and Harvey Wheeler, 1962). It even resonated with a popular nonfiction book of the period that imagined the coming apocalypse in ideological rather than military terms, C. S. Lewis's *The Abolition of Man* (1943), whose title interestingly anticipates Orwell's working title for *1984,* "The Last Man."

Michael Anderson's *1984* similarly emphasizes the political relevance of Orwell's vision. A joint U.S.-UK production, the film connects itself to the two ongoing cinematic series then exploring, in different ways, the more general sense that "the dreadful is still about to happen," as well as the more particular threat posed to Western society by Soviet communism, which in these films (as in the popular imagination) figures chiefly as a totalizing, dehumanizing reorganization of human experience (not as a form of liberation from the prison house of social class or a change of governance over the means of production that delivers society to greater economic equality). These two cinematic series are the science fiction film and the film noir (both popularly produced and consumed on both sides of the Atlantic). American and British science fiction films of the 1950s normally feature a more or less conventional happy ending, with the threat to human life and social stability defeated by the aroused forces of the state, which, in extremis, manages to contrive with the assistance of science some means of wiping out ants and grasshoppers turned into ravening giants by exposure to radiation or destroying aliens bent on the conquest of Earth by discovering some weakness in their seemingly invulnerable posthuman bodies.

The film noir, in both its British and American versions, offers a more pessimistic, archly secular view of modern society trapped in the nightmare of the dark city, which is home to the anomic, the pathological, the destructively self-indulgent, and the psychopathically criminal. The helpless, hopeless protagonist of film noir customarily finds himself thwarted and destroyed by forces beyond his own control that also elude his full understanding, including those to be found within his own character. Typical is *D.O.A.* (Rudolph Maté, 1950), in which Edmond O'Brien, at that time a veteran of noir films (such as *The Killers* [1946] and *An Act of Murder* [1948]), plays a man who resists a life of full commitment and responsibility, only to discover too late the virtues of respectability as well as the comforts of a loving relationship; poisoned for motives he only dimly understands in the end, he lives long enough to bring his killers to rough justice and to confess the bizarre tale of his own murder to the police.

Michael Anderson's *1984* offered O'Brien, who plays Orwell's protagonist Winston Smith, the opportunity to reprise this noirish role as a man doomed from the outset despite his best efforts to master destiny and circumstance; O'Brien is paired with Jan Sterling, whose Hollywood career had also taken full advantage of her ability to play the vaguely dissatisfied, erotically charged, and self-destructive woman of film noir (as in *Female on the Beach* [1955] and *The Human Jungle* [1954]). The novel's Julia can readily be fitted into the generic noir role of what some critics call the *femme attrapée* (the trapped woman).

The film's opening sequences evoke the similar yet divergent threats posed to human life and happiness by "invader" science fiction and the film noir, signaling a significant departure from the novel, which, the credits affirm, has been "freely adapted." Orwell, in contrast, begins with the movement in Winston's consciousness that leads him to the rebellious act of writing a diary. Anderson emphasizes spectacle rather than character. A huge explosion fills the screen, as a narrator recounts that the "atomic wars of 1965" resulted in a world dominated by three large superstates, one of which, Oceania, incorporates the former British Empire. Thereupon follows an airplane shot (using very accurate and detailed models) of modern London, delineating a skyline dominated not only by familiar landmarks (Big Ben, the Tower Bridge) but by four archly modernist towers, eventually revealed to be the ministries of the Ingsoc establishment.

Streets, strangely drab and vacant, come into view as sirens warn of an imminent rocket attack and pedestrians, including Winston, duck for cover, he in a storefront where Julia, hitherto unknown to him, also takes shelter. The warnings are accompanied by further establishing shots of a city whose streets are lined with houses that are obviously the as yet unrepaired sites of previous explosions. Barricades of barbed wire are everywhere in evidence, suggesting the anticipation of some imminent invasion, to be opposed, it seems, by the troops of motorcycle police, who, dressed in fascist-style garb, ride out in perfect geometrical formation to assert the power of the state. The apocalyptic futurity evoked in these sequences recalls the similar atmosphere created in the "creature features" and "spacemen" science fiction films of the era.

Anderson's London, however, is not only the frightening world of the future; it is also the dark cityscape conjured up by the modernist id. A bit later in the film, Winston finds himself walking along the rain-swept nighttime streets of the otherwise deserted "people's quarter," in violation of the rules governing the behavior of Outer Party members, who are forbidden to associate with the "proles." Spotting a police patrol van creeping along the street behind him, Winston tries to act nonchalantly, but his deep fear of the authorities is evident. Questioned by the patrol, he offers an answer so unsatisfactory that he is summoned the next day to report to officials in his own ministry, and here he narrowly averts a trip to the Ministry of Love, where, we learned earlier, those who deviate from the principles of Ingsoc are "re-educated." His savior is the Inner Party official O'Connor (O'Brien in the novel), who begins to befriend him but, it will turn out, ironically constitutes the worst threat to Winston's individuality because he later takes charge, using terrible pain and unendurable fear as his tools, of showing his ostensible friend the error of his ways. The paranoia that often grips the noir protagonist, suggesting the dark side of the everyday and the apparently benign, here reveals its underlying grain of truth. Everyone in his world is in fact out to "get" Winston, including the only person in his world besides Julia who takes any interest in him.

Anderson's *1984* constructs a future characterized above all else by a double threat—from unseen and distant enemies launching rocket attacks and from the ostensible forces of law and order, who are inalterably opposed to individual happiness. As O'Connor later explains, the threat from without is as carefully engineered as the threat from within: the "foreign attacks" are launched by the state against its own people. The film's preface suggests how viewers are to understand this world marked by domestic and international insecurity: "This is a story of the future—not the future of space ships and men from other planets—but the immediate future." In Orwell's novel, as I remarked earlier, it is easy to forget that the present moment of 1984 marks only the near completion of the totalizing processes set in motion by socialist revolution. The petty rebellion of Winston and Julia results not only from the failure of the Anti-Sex League to conquer the orgasm and eliminate the fact of romantic relationships but also from the inexorable but glacial progress toward a version of Newspeak that will make it impossible to commit the thoughtcrime of which they are self-confessedly guilty.

The film's statement that it limns a portrait of "the immediate future" suggests that the trends that have developed to their horrifying fullness in that future are already present; hence its engagement with the terrors, as codified by the generic plenitude of commercial filmmaking, of nuclear apocalypse, alien invasion, and the dark city. Our world thus finds an interesting correlative in the postulated future, where Winston and Julia are terrified by the same fears and engaged by the same desires, and where it is also true that "the dreadful is still about to happen." The film thus gives voice to the structure of feeling that dominated social and political life in the West during the 1950s.

In Anderson's film, as in Orwell's novel, history, understood as the unfolding of the order of things, has not yet come to an end, a point made tellingly by art design and the way in which mise-en-scène is deployed. For the London renamed Airstrip One is both the London of our present, its major monuments intact, its familiar outlines still readable from the air, and yet also a location significantly transformed by a new order, with its ostentatiously modern buildings, flaunting the fascist minimalist style, standing for the transformation of English society otherwise effected by ideological rather than material means. Within the film, the interior spaces correspond precisely to the now prevailing class order, with the proles inhabiting and working in uniformly old and crumbling houses and tenements of apparent Victorian provenance, the Inner Party members living in blocks of shabby, newer flats but working in the ultramodern ministries (dominated by interiors of glass, shiny chrome, and austere interior decoration), and the Outer Party members enjoying spacious apartments appointed in the somewhat old-fashioned style of an exclusive gentleman's club (heavy wood furniture, thick carpets, paintings on the wall, subdued lighting, and tasteful objets d'art, an environment devoted to bachelor comfort that is lovingly policed by obsequious, white-coated servants).

Appropriately, it is the public world of *1984* where the aesthetic and technical transformation of modernity is most in evidence. The foyer where the Ministry of Love employees conduct their daily "two-minute hate" sessions is dominated by a futuristic telescreen, slick tiled floors, and minimalist furnishings. When Winston and Julia are finally apprehended by the Thought Police, they are hustled off to the same ministry, there to be tormented and corrected in laboratory-like surroundings that strongly evoke the modern hospital's operating theater. All steel fittings and built-in cabinetry that almost exudes the smell of sterility, this filmic manifestation of the infamous

Room 101 seems a fit environment for the redesign of the human spirit, including and especially the expunging of any and all connections to the not so technically advanced, ideologically suspect past. Orwell's dystopic technophobia here finds a convincing visual correlative. Never to be fully incorporated into the society for which they supply the needed labor, the proles, in contrast, are allowed to live in a kind of timelessness. To judge by their dress, demeanor, and social habits, they inhabit, and will always inhabit, a frozen past moment, seemingly in the 1930s. Their world is untransformed by the technological and ideological "advances" of the present; their past, we might say, can neither tend toward nor imagine any future.

Talking to the Past: *1984* (1984)

In the novel, Orwell strongly suggests that what Inner Party members fear most is the threat they imagine is posed by the past, the time when things were different. The past is memorialized in written records, most especially in that most English of institutions, the *Times*, whose daily issues provide an accurate account of what is happening (or, more accurately, what the Inner Party wishes that others in Oceania think is happening). Orwell, somewhat comically, seemingly cannot imagine a future from which that most middle-class of modern institutions, the daily newspaper, has disappeared. And hence the *Times*' morgue, a self-perpetuating archive, threatens the ideological seamlessness of the present, containing, if left unrevised, the truths that might be used to debunk what the Party at any present moment claims is, was, and always has been.

Along with a host of others, Winston is employed in correcting the past and thereby turning the printed "record" into irrefutable proof of the regime's continuing economic and social progress (measured by Oceania's supposed unalloyed success in meeting or surpassing the goals of successive multiyear plans, a falsified account of collective accomplishments providing what we now term *metrics*). The Inner Party, of course, fears not only the facticity of the historical record but something more dangerous: the historical consciousness, another aspect of modernity that, like the *Times*, Orwell cannot imagine that the future might find rendered irrelevant—absent, that is, the most extreme and constant exercise of revisionism. In fact, in a famous formulation, Ingsoc proclaims that the project of totalization can go forward only through the effective engineering of the historical consciousness: "Who controls the present, controls the past. Who controls the past, controls the future." The flux and flow of historical accident is transformed into the never-deviating path of material and social progress that constitutes the master narrative contrived for Oceania by the Party. Enlightenment thought seeks in this way its destined fulfillment, with the Inner Party's reconstructed history eliminating the irregularities and inconsistencies of actual events in a gesture that confirms the triumph of totalization and the application of reason to nature and human experience.

How different, according to the theorists of postmodernity, like Fredric Jameson, is the contemporary experience of temporality. We live, Jameson laments, in a world marked by "the failure of the new" and "imprisonment in the past." The craze for nostalgia films (in which we return to the cultural past, there "to live its strange old aesthetic artifacts through once again") marks, for Jameson, our collective inability "to focus on our own current experience." Such an artistic turn toward a pastness whose dominant form is pastiche (the confection of new objects from materials borrowed from some repertoire of dead styles) is nothing less than "an alarming and pathological symptom of a society that has become incapable of dealing with time and history."[18] The explanation is not far to seek, as Jameson suggests: "No society has ever been so standardized as this one, and . . . the stream of human social and historical temporality has never flowed quite so homogenously." Our sense that time flows smoothly, paradoxically enough, depends, he suggests, on the fact that "where everything . . . submits to the perpetual change of fashion and media image . . . nothing can change any longer."[19] History, it seems, has ended, and because meaningful, substantial change is no longer possible, ideological conflict no longer matters.

This contemporary sense that there is no longer a future to fear or hope for (that what is to come is only a constant, thorough flux of fashion) finds a reflex, I think, in Radford's general approach to adapting Orwell's novel, which was, he is reported as saying, "not a futuristic fantasy but a satire on his own world, an extreme vision of Britain in 1948 at the height of the Cold War."[20] In rejecting *1984* as a futuristic fantasy, Radford, of course, was aligning himself with a certain, minority school of Orwell criticism, most notably represented by Anthony Burgess's extended essay *1985*. The resulting film, however, is hardly a realistic treatment of Clement Atlee's "Austerity Britain"; Radford does not offer a richly detailed mise-en-scène meant to evoke the postwar era, especially the dreadful year of 1948. Britain as such, in fact, despite what Radford says, is only barely a part of this "vision."

Dominating the director's aesthetic, instead, is a form of abstractionism that yields a fantasy with a somewhat different point of departure; this is apparently what Radford means when he says that the book is actually "an *extreme* vision of Britain in 1948." Not the future, as Orwell might imagine it as a nightmare, but an alternative present in which the dreadful has already happened. Such a present, delivered to the notion of the dreadful, has been emptied of its specificity. Radford, that is, rejects any concerted attempt to evoke the actual past through the accumulation of authentic detailing. The film becomes, therefore, a bad dream of the past, one that viewers of our present can contemplate but not share.

We can, in other words, regard Radford's *1984* only as a museum piece; it allows us to visit from a now palpably unbridgeable passage of time a bygone era of material privation and the momentous conflict of worldviews. We look only backward at an era saturated with the importance of history. But we cannot connect to that evoked past moment through the workings of illusionism; the postwar era cannot find a real space in the film, cannot come to life again for us. The dystopian projection of Orwell's fiction, deprived of its ability to evoke a "possible historical trajectory" toward the future, becomes instead the stuff of a museum piece, a cultural artifact still so valued it merits reverential display but so irrelevant to the present moment that the process of representation can no longer be controlled by mimesis (the imitation of the real). Instead, in a kind of anti-Menippean gesture, a thoroughgoing abstractionism reduces the novel's set of ideas to little more than reminiscences about a now long-silenced political dispute.

To put this another way, the past we are invited into is the literary past (the fact of Orwell's novel, not its "message"). What Radford intends for us to do in regard to that past is nothing more than, as Jameson sardonically puts it, to "live its strange old aesthetic artifacts through once again." Futurism, the sense that society is evolving both technically and politically, is not to be found in any image of Radford's film; its settings are uniformly shabby, grim, and underdecorated, leaving no space for Orwell's technophobia. There are no material, visible signs of an evolving world. When the mise-en-scène requires some deviation from the presumed technical developments of 1948, these "innovations" take shape only as those in the past might have imagined them as being. The film's telescreens, for example, dimly lit, clumsy-looking, and old-fashioned tube appliances, make us aware, more than anything else, of the limitations of the novelist's imagination, of his inability to conceive what the future might truly hold. We are a long way from Anderson's claim that his film represents the "immediate future."

Perhaps Radford's decision to debunk the conventional, majority reading of *1984* as dystopian fantasy reflects not only an interpretive decision
(based on his doubtful conclusions about the novelist's intentions, gleaned
from interviews with his family). It may reflect as well his sense that the
book's political theme (its "futuristic fantasy") no longer has any relevance
for the present, a view shared by some, but by no means the majority of,
Orwellians in the twenty-first century. It may also reflect, I believe does
reflect, his inability as an artist to conceive any way that the book might be
updated, for this would surely involve conceiving of some future moment we
now collectively desire and fear that might correspond to the same element
in *1984*. And such a future is no longer available to be represented.

Three decades earlier, Anderson found it possible to connect his staging
of the novel to the inchoate paranoia and disestablishmentarianism of film
noir, to the fear of invasion from without and "takeover" from within that
makes science fiction of the age such an interesting symptomatic response
to Cold War politics. But in Radford's version, the past moment represented
in the film leads only toward a yet more distant past, as well as toward a
general sense of dissolution and decay, marking the ostensible futurity of its
reference as an irrelevant nightmare. It is now no more than the dead-ended
vision of what was to come that took shape in an artist disillusioned in equal
proportions by the excesses of Stalinism and the failures of the Labor Party
in his own country to produce the fruits of wartime victory, including social
equality and material comforts.

Anderson was influenced by the two cinematic series that in his own
time constituted an affective framework that would allow him to transfer
the structure of feeling in Orwell's novel to the screen; the science fiction
film and the film noir both conjured up and connected to strong experiences of threat and dread. Radford, in contrast, imagines his version within
the series that most characterizes the British film industry of the era, what
Andrew Higson has identified as the heritage film.[21] Superficially, *1984* seems
to constitute itself as a kind of anti–heritage film, if we conceive that this
genre, best known through the reverential period adaptations of nineteenth-
and early-twentieth-century British fiction produced by Merchant Ivory, as
generally dominated by a luscious pictorialism that especially evokes the
grandeur and opulence of life in a bygone era among the aristocracy and
haute bourgeoisie.

But Radford's film in its own way conforms to what Higson calls the
"pictorialist museum aesthetic," with its carefully stylized, beautifully pho-

tographed images of drably dressed characters and a London evoked met-
onymically through bombsites, ramshackle apartment buildings, railways in
a state of ill repair, and a deserted country forest setting of deepest, almost
supernatural green (the Edenic location for the initial assignation of Julia
and Winston as well as the image in Winston's mind since childhood of a
place beyond conflict and strife).[22] But this pictorialism, enhanced by a rich,
original sound track, is abstract in the expressionist manner, with every image carefully loaded with both tone (Stimmung) and a sense of psychological
rather than forensic reality. The well-known monuments of this city do not
figure for the most part in its visual program; they are evoked solely (and
tellingly) in a vague, faded lithograph of what appears to be St. Clement's.
The city "we know" is present only in an image that is barely readable, appropriately marked as culturally irrelevant (it is an item for sale in a junk
store). The ruins and drab interiors of the film's Airstrip One, moreover,
seem to hearken back more to late-Victorian than late-1940s Britain (this
is especially true of the sequences that depict the proletarian quarters).
Radford's is thus a pictorialism in the style of Lang's classic, studio-bound
version of the future, *Metropolis* (1927), rather than that of David Lean's *A
Passage to India* (1984). His *1984* is not futuristic, of course, in the sense
that *Metropolis* certainly is; it is instead fantastic, as stated before, the bad
dream that never came true, the alternative past.

While Orwell's political ideas are duly evoked in the film, Radford shows
no real interest in exploring them in depth. The dramatization of the workings of doublethink that is the centerpiece of the dialogue between O'Brien
and Winston is particularly weak, with the material from the novel drastically abbreviated and decontextualized. But heritage films typically do not
develop themes in depth. Instead, in the manner of the European art film, the
heritage film characteristically offers narratives that are "slow-moving, epi-
sodic, and de-dramatized," with an emphasis on "ensemble performance."[23]
They are films in which style is more important than content. These are the
qualities most evident in Radford's version of the novel, to be explained in
part by his belief that the book was deficient, being nothing more than a
political essay animated by "cardboard characters."[24] In heritage films, as in
Radford's *1984*, the adaptation process generally creates a "space in which
character, place, atmosphere, and milieu can be explored." Such an aesthetic
is the polar opposite of the Menippean satire of Orwell's novel, a form that,
to quote Northrop Frye again, concerns itself "less with people than with
mental attitudes . . . with abstract ideas and theories."[25]

An Honored Object?

Considering the relevance of Jameson's theories of the postmodern to the advent and flourishing of the heritage film since the 1980s in Britain, Higson remarks that the evoked worlds of such films seem, on the one hand, to be "cut off from the 'real' historical past" and, on the other hand, "to have severed [their] connections with the present." We might call this effect museumizing, a process of hermetic sealing that preserves the object even as it renders it profoundly irrelevant. Higson explores the irony implicit in such an "aesthetic" orientation toward the rediscovery of the past and the artistic objects of the past: "The strength of the pastiche in effect imprisons the qualities of the past, holding them in place as something to be gazed at from a reverential distance, refusing the possibility of a dialogue or confrontation with the present."[26] If dystopian fiction, as suggested earlier, must address the fears and desires of the present as they might shape a future of startling alterity, then as a form it is no longer available in postmodernity. Radford's adaptation of *1984*, more than anything else, represents the sense in which the novel has now been respectfully imprisoned, becoming an honored object "to be gazed at from a reverential distance."

In the heritage film based, as most are, on a literary original, Higson tells us, "the source text is as much on display as the past it seems to reproduce," the art object substituting, in a deeply postmodern fashion, for anything it might be conceived as representing.[27] In this respect, most revealing was Radford's decision to cast, as Winston Smith, an actor (John Hurt) who at the time of the film's making bore an uncanny resemblance to the tubercular, painfully thin, and wasted author. Radford has acknowledged that Orwell's *1984* is in a profound sense a political essay, the most influential text written by twentieth-century Britain's most politically engaged novelist.

Yet, by abandoning or minimizing what the novelist had to say, his film memorializes Orwell not as a polemicist, warning of a future that might in some sense be in the offing, but as a character trapped within his own bad dream of his present, speaking only to the dead (among whom he includes himself) and to a bygone era, author of a diary of discontent that nobody but his enemy will ever read. Radford begins with the novel's historicist credo: "Who controls the present, controls the past. Who controls the past, controls the future." Is this Orwell's truth? Is it his warning? The film contends that it is neither—at least no longer. Radford's epigraph, in an ironic postmodern gesture, not only evokes absence (the Enlightenment paradigm of progress, now "lost" to us) but also voices a profound, perhaps bitter nostalgia for the palpable sense of a future that the form of futurist fantasy cannot conceptualize now that history seems to have ended, yet still finds it possible, even necessary, to name.

Notes

The epigraph to this essay is drawn from Anthony Burgess, *1985* (Boston: Little, Brown, 1978), 29.

1. Frye quoted in Gorman Beauchamp, "*1984:* Oceania as an Ideal State," *College Literature* 40, no. 1 (1984): 2. Much the same point about Orwell's book is made by Irving Howe, who opines that it is "mostly Menippean satire." Irving Howe, "*1984:* Enigmas of Power," in *"1984" Revisited: Totalitarianism in Our Century,* ed. Irving Howe (New York: Harper and Row, 1983), 8.

2. That a political reading of the novel has dominated its reception is indicated by comments such as those of Raymond Williams, who observes, "*Animal Farm* and *Nineteen Eighty-Four* lead us inevitably to the controversial history of the Russian revolution and its consequences, and to the politics of the cold war." Raymond Williams, introduction to *George Orwell: A Collection of Critical Essays,* ed. Raymond Williams (Englewood Cliffs, NJ: Prentice-Hall, 1974), 2. Tellingly, the 1955 film adaptation of *Animal Farm* was funded covertly by the CIA; the agency also oversaw the film's distribution. See John Rodden, *Scenes from an Afterlife: The Legacy of George Orwell* (Wilmington, DE: ISI Books, 2003), 45.

3. See, for example, William Steinhoff's comment that "whoever investigates the background of *1984* must pay particular attention to the work of James Burnham." William Steinhoff, *George Orwell and the Origins of "1984"* (Ann Arbor: University of Michigan Press, [1975]), 43. An in-depth examination of the pervasive influence of Burnham is provided by R. B. Reaves, "Orwell's 'Second Thoughts on James Burnham' and *1984,*" *College Literature* 40, no. 1 (1984): 13–21. The "totalitarian" approach to the novel is interestingly exemplified by the celebratory collection of essays edited by Irving Howe, *"1984" Revisited.* See particularly Michael Walzer's "On Failed Totalitarianism," which reads the novel in the tradition of Arthur Koestler's *Darkness at Noon,* as a thinly disguised portrayal of the Stalinist Soviet Union.

4. Orwell quoted in Steinhoff, *George Orwell,* 54.

5. Ibid.

6. Paul Ricoeur, "Ideology and Utopia as Cultural Imagination," in *Being Human in a Technological Age,* ed. Donald M. Borchert and David Stewart (Athens: Ohio University Press, 1979), 121–22.

7. Lewis Mumford, *The Story of Utopias* (New York: Viking Press, 1962), 7.

8. Burgess, *1985,* 18.

9. J. P. Telotte, *Science Fiction Film* (Cambridge: Cambridge University Press, 2001), 125–26.

10. Ibid.

11. Burgess, *1985,* 18.

12. Telotte, *Science Fiction Film,* 127.

13. Ibid.

14. Terry Eagleton, *The Illusions of Postmodernism* (Oxford: Blackwell, 1996), 16.

15. Harold Bloom, introduction to *George Orwell's "1984"* (New York: Chelsea House, 1987), 4.

16. Walzer, "On Failed Totalitarianism," 103–21.

17. Rodden, *Scenes from an Afterlife.*

18. Fredric Jameson, "Postmodernism and Consumer Society," in *The Cultural Turn: Selected Writings on the Postmodern, 1983–1998,* ed. Fredric Jameson (London: Verso, 1998), 6–10.

19. Fredric Jameson, "The Antinomies of Postmodernity," in Jameson, *Cultural Turn,* 59.

20. Radford quoted in Michael Billington, "A Director's Vision of Orwell's *1984* Draws Inspiration from 1948," *New York Times,* June 3, 1984, 27.

21. Andrew Higson, *English Heritage, English Cinema: Costume Drama since 1980* (Oxford: Oxford University Press, 2003).

22. Ibid., 39.

23. Ibid., 26.

24. Radford quoted in Billington, "Director's Vision," 27.

25. Frye quoted in Beauchamp, "Oceania as an Ideal State," 2.

26. Higson, *English Heritage,* 65.

27. Ibid., 20.

DISENCHANTMENT AND REBELLION IN *ALPHAVILLE*

Alan Woolfolk

There are other creative sensibilities besides the seriousness (both tragic and comic) of high culture and of the high style of evaluating people. . . . For instance, there is the kind of seriousness whose trademark is anguish, cruelty, derangement. Here we do accept a disparity between intention and result. I am speaking, obviously, of a style of personal existence as well as of a style in art; but the examples had best come from art. Think of Bosch, Sade, Rimbaud, Jarry, Kafka, Artaud, think of most of the important works of art of the 20th century, that is, art whose goal is not that of creating harmonies but of overstraining the medium and introducing more and more violent, and unresolvable, subject-matter. . . . And third among the great creative sensibilities is Camp: the sensibility of failed seriousness, of the theatricalization of experience. Camp refuses both the harmonies of traditional seriousness, and the risks of fully identifying with extreme states of feeling.

—Susan Sontag, "Notes on 'Camp'"

And what of pure poetry? Poetry's absolute power will purify men, all men. "Poetry must be made by all. Not by one." So said Lautréamont. All the ivory towers will be demolished, all speech will be holy, and, having at last come into the reality which is his, man will need only to shut his eyes to see the gates of wonder opening.

—Paul Éluard, "Poetic Evidence"

Jean-Luc Godard's *Alphaville, une étrange aventure de Lemmy Caution* (1965) appears on first viewing to be a typical variation of a classic science fiction film in which humanity has been invaded and is threatened with complete colonization by an alien force—in this case an alien computer, Alpha 60, which rules in the name of scientific logic. Indeed, Godard's film even features a deranged scientist-villain, Leonard Von Braun (Howard Vernon), as the chief human agent of this alien colonization who has himself been converted into the apparently perfect scientific-technocratic man without a trace of human emotion. However, the deeper theme of *Alphaville*, as in nearly all science fiction films, appears to be not merely science but a disaster of unimaginable proportions that is nonetheless imagined.

As Susan Sontag wrote in 1965, the same year that *Alphaville* was produced and released, the defining motif of science fiction films is the imagination of disaster. "Science fiction films are not about science. They are about disaster. . . . In science fiction films disaster is rarely viewed intensively; it is always extensive. It is a matter of quantity and ingenuity." In the typical science fiction film, as opposed to novel, the extensiveness of the disaster is normally conveyed in a direct and immediate manner through images and sounds, rather than words, which "have to be translated by the imagination" for effect.[1] Even the most thoughtful science fiction films (e.g., *Things to Come* [1936], *The Day the Earth Stood Still* [1951], *Invasion of the Body Snatchers* [1956], *Planet of the Apes* [1968]) tend to rely heavily on nonverbal communication and to exhibit many of the characteristics of a spectacle. *Alphaville* is unusual in this regard: it is not a spectacle and yet it relies to a significant degree on nonverbal communication. The images and sounds of the film have to be translated or interpreted by the viewer, while the impoverishment of language directly reflects the extensiveness and ingenuity of a disaster that is clearly inward and spiritual. The sensuous expressions of Godard's film demand intellectual work, while the language of the dystopian city of Alphaville is at risk of becoming a toolkit of functional devices, which require little or no interpretation, for manipulating the population. The truncation of language has resulted in a situation in which evocative, symbol-laden images must be employed both to define the nature of the disaster and to fight off the poverty of words, in order to create the possibility of a new unity of image and language necessary for ultimately overcoming the disaster.[2]

The Architecture of Disenchanted Space

Godard uses a variety of evocative images centered on the theme of dehumanization or, more precisely, spiritual disenchantment and desecration to communicate the essential nature of Alphaville's disaster: robotic human behavior, rule by artificial intelligence, ominous signage (e.g., identifying numbers tattooed on human bodies), bizarre "swimming pool" executions of spiritual dissidents, featureless functional architecture circa 1960, and, perhaps most important, an array of chiaroscuro and monochrome scenes, which borrow heavily from German expressionism and film noir, to reflect the inner darkness of this world. Of course, we are *told* that those who do not succumb to the logic of Alphaville ultimately sink into despair and are at risk of committing suicide. But the death agony of secret agent Henri Dickson (Akim Tamiroff) conveys far more powerfully than words the self-destructive despair of those who cannot be assimilated: he is the perfect image of knowing dissipation as he expires in the embrace of a state seductress, bequeathing in his death throes a copy of Paul Éluard's *Capitale de la douleur* (1926; *Capital of Pain*) to his successor, Lemmy Caution (Eddie Constantine).

On the surface, *Alphaville* appears to be "a straightforwardly dystopian vision of technocratic progress, a sci-fi noir." But, as Chris Darke argues, rather than proposing that the future is "latent in the present," *Alphaville* dares to assert that the future is "already taking place." Employing "a short-circuited kind of realism that is applied to the least likely of genres, science fiction, and that serves to make strange a reality that is seen as already being so strange that it positively demands the strategy," Godard filmed some of the newest buildings and structures in Paris constructed in the late 1950s and early 1960s, including the Maison de l'ORTF, the Esso Tower in La Défense, and high-rise housing blocks. As Darke points out (drawing upon the work of Marc Augé[3] and Peter Wollen), Godard constructs a city that "is almost entirely made up of architectural non-places: the city is a patchwork of transitional zones—corridors, staircases, offices, hotel rooms—liberally interspersed with their characteristic signage—arrows, numbers, neon."[4] Constructing such a city of disenchanted nonplaces, of space that is simply traveled through, is an effective way to depict a negative utopia, since the word *utopia* of course means "not a place" or "no place." But Godard takes the unusual approach of suggesting that this city of nonplaces already exists in the present and furthermore that it is located in Paris, a veritable city of specific places rich with historical texture and meaning, grand public monuments, and exquisite personal spaces.

Peter Wollen argues that films that focus on the construction of narrative space, as opposed to those that subordinate narrative to spectacle, attempt to establish stable inhabitable places, to pay attention to the location of particular characters in particular places.[5] In Alphaville, this construction of narrative space cannot take place; indeed it is in effect forbidden by the very nature of the regime. The establishment of place can take place only outside Alphaville and the locale of the film in "the exterior" or outlands. This would not pose a problem if *Alphaville* were simply another science fiction spectacle employing a series of nonplaces to be looked at as settings for disaster. However, one of the critical symptoms of Alphaville's spiritual disaster is the omnipresent architecture of nonplaces, of empty transitional spaces devoid of any public or private significance. They reflect both the lack of stable personal identities among the inhabitants of Alphaville and the inability to form any, for without an established architecture of significant places, a coherent narrative of personal character development becomes, to say the least, problematic.

Alphaville's omnipresent architecture of nonplaces invites further analysis in light of Vivian Sobchack's thesis that film noir is in fact defined by the absence of stable domestic spaces. In her effort to move beyond the question of whether film noir is a style or a genre, Sobchack employs Russian theorist Mikhail Bakhtin's concept of chronotopes "to locate and ground that heterogeneous and ambiguous cinematic grouping called film noir in its contemporaneous social context," specifically the unsettled, contingent world of post–World War II American society: "The hotel or boarding house room, the cocktail lounge, the nightclub, the diner or roadside café, the bar and roadhouse, the cheap motel—these are the recurrent and ubiquitous spaces of film noir that, unlike the mythic sites of home and home front, are actual common-places in wartime and postwar American culture. Cinematically concretized and foregrounded, they both constitute and circumscribe the temporal possibilities and life-world of the characters who are constrained by them—and they provide the grounding premises for that cinematic grouping we have come to recognize as noir."[6] For Bakhtin, a chronotope "expresses the inseparability of space and time (time as a fourth dimension of space)." In fact, "it is precisely the chronotope that defines genre and generic distinctions, for in literature the primary category in the chronotope is time."[7] In adapting Bakhtin's concept, Sobchack stipulates the inseparability of space and time in film noir in terms of the chronotope of "lounge time," with impersonal, transitional, nondomestic locales, such as nightclubs and hotel rooms, representing its spatiotemporal reality in film. But, as R. Barton Palmer has argued, Sobchack theorizes the spatial dimension of film noir at "the wrong level of specificity" while practically ignoring the temporal dimension altogether. "The problem with Sobchack's anatomy of noir premises, to state it simply, is that in too many noir films the main settings are not cocktail lounges, cheap bars, bus stations, and roadside diners." The chronotope of lounge time is misconceived, with the spatial dimension defined too narrowly and the temporal dimension barely if at all.[8]

This does not mean that Sobchack's analysis is completely without merit, for by focusing on the architecture of nonplaces she helps to highlight an important spatial dimension of film noir, however incomplete, that is present in *Alphaville,* while at the same time inadvertently drawing attention to the primary importance of the temporal. As Palmer explains, Bakhtin's concept of chronotope gives priority to time, not space: film noir defines a narrative structure, a form of cinematic modernism, that emphasizes the "dark pasts" of its protagonists. And it is the contingent vulnerability of present life to the dark past that determines the transitional, discontinuous spatial dimension of the noir world and the unstable, threatened identities of noir protagonists.[9]

The Disenchanted Present

Alphaville reverses the logic of the noir world by linking the denial and destruction of the past to the architecture of nonplaces and the vulnerability of personal identity. More specifically, *Alphaville*'s abstract, scientific conception of time as a series of divisible, repeatable units seems to be behind the conclusion that the present is all that counts temporally, spatially, and personally. "No one has lived in the past and no one will live in the future," proclaims Alpha 60. "The present is the form of all life." Consequently, "time is like a circle which is endlessly described. The declining arc is the past. The inclining arc is the future." Here, Godard seems to operate from the Bergsonian premise that it is our reduction of the single, indivisible quality of time, the *durée,* to a series of discrete units in space that disrupts our genuine experience of time and memory. On the one hand, "pure duration is the form which the succession of our conscious states assumes when our ego lets itself *live,* when it refrains from separating its present state from its former states."[10] According to Bergson, our fundamental self is ultimately identical with the experience of pure durée. On the other hand, our surface self is associated with the practical nature of intelligence, with survival and progress in technical skills, which Alphaville represents to the point of having completely suppressed the experience of the fundamental self. Consequently, Lemmy Caution appears to represent a threat to the Alphaville regime precisely because he insists upon the priority of the fundamental self, when he proclaims that he "believes in the immediate data of *la conscience,*" which is best translated as "consciousness" to capture the Bergsonian implications.

The critique of contemporary society contained in *Alphaville* parallels and perhaps draws upon the Frankfurt School's critique of mass culture. In 1942, Max Horkheimer wrote in criticism of what he would later call "the culture industry" (using the truncation of time in film as his primary example) that "this trimming of an existence into some futile moments which can be characterized schematically symbolizes the dissolution of humanity into elements of administration." He continues, making explicit the Bergsonian implications: "Mass culture in its different branches reflects the fact that the human being is cheated out of his own entity which Bergson so justly called '*durée*.'"[11] Of course, the French new wave cinema opposed the rise of the culture industry and the role of film in it. From the notion that directors are auteurs conceiving and creating personal films to the praise of low-budget films made outside the system, the French new wave movement defined itself in opposition to the established big-budget spectacles of the French film industry.[12] As one of the pivotal figures in this movement, Godard went one step further in *Alphaville,* demonstrating that film itself may be self-consciously employed in a highly theoretical manner to reassert the primacy of durée.

The spiritual catastrophe of Alphaville appears to be a direct result of what German social theorist Max Weber in his seminal essay "Science as a Vocation" (1919) calls the process of the intellectualization and rationalization of the world, of which "scientific progress is a fraction, the most important fraction." Weber tells us that this process does not lead to "an increased and general knowledge of the conditions under which one lives." Rather, it eliminates meaning from collective and individual life, leaving only the instrumental priorities of the present moment: "It means something else, namely the knowledge or belief that if one but wished one *could* learn it at any time. Hence, it means that principally there are no mysterious incalculable forces that come into play, but rather that one can, in principle, master all things by calculation. This means that the world is disenchanted." For Weber "this process of disenchantment . . . this 'progress,' to which science belongs as a link and motive force," raises the question, Are there "any meanings that go beyond the purely practical and technical?" Inevitably,

Weber argues, disenchanted modern culture must confront the conclusion of Tolstoy that "for civilized man death has no meaning . . . because the individual life of civilized man, placed into an infinite 'progress,' according to its own imminent meaning should never come to an end." The progressive assumptions built into modern civilized life give both life and death the stamp of meaninglessness.[13]

Nonetheless, the disaster of scientific progress that Alphaville exemplifies may also open up new spiritual opportunities and challenges. As philosopher Charles Taylor has explained with great insight, our lives unfold and take on meaning against preexisting horizons of significance that are beyond the individual's choice and that can only be suppressed or denied in self-defeating moves, which have become all too prevalent in a "subjectivist civilization" characterized by "weak" or instrumental evaluations. Such an "inescapable framework" normally defines a background of intelligibility that makes "qualitative distinctions" between the high and the low, the good and the bad, the dignified and the undignified—"to think, feel, judge within such a framework is to function with the sense that some action, or mode of life, or mode of feeling is incomparably higher than others which are more readily available to us." These frameworks have always defined human lives, even when the frameworks have undergone transformations, been challenged, or grown increasingly unintelligible, as they have in modernity, and especially the twentieth century. "What Weber called 'disenchantment,' the dissipation of our sense of the cosmos as a meaningful order," Taylor argues, "has allegedly destroyed the horizons in which people previously lived their spiritual lives." But Taylor maintains that the very lack of a preexisting, unchallengeable framework has created a new and very different type of spiritual agenda with its own spiritual obstacles and risks that cannot be denied.[14]

Taylor contends that our cultural horizons, our spiritual frameworks, have become "problematic" in the modern world, that the "existential predicament" on our "spiritual agenda" is no longer one in which "an unchallengeable framework makes imperious demands which we fear being unable to meet." Rather "the form of danger . . . which threatens the modern seeker . . . is something close to the opposite: the world loses altogether its spiritual contour, nothing is worth doing, the fear is of a terrifying emptiness, a kind of vertigo, or even a fracturing of our world and body-space." However, the very fact that we no longer have an established background of intelligibility has itself been taken up as the basis of a new kind of framework—one in which the model of a higher life "consists precisely in facing a disenchanted universe with courage and lucidity." Within this disenchanted horizon, dignity comes from the "ability to stand unconsoled and uncowed in face of the indifferent immensity of the world" and to find purpose in confronting it.[15] Weber clearly belonged within this spiritual framework, as did Nietzsche, as did Camus, as have so many twentieth-century spiritual protagonists in life and art. Godard's Caution must also be added to these ranks. Unlike noir protagonists, Caution exhibits sensitivity and a response to the horizon of disenchantment (witness his Pascalian allusion: "the silence of these infinite spaces frightens me") that place him in a line of spiritual descent that has insisted upon finding dignity and higher purpose where others have found distraction and ruin.

Strategies of Rebellion: Camp and Poetic

Godard's central image of resistance to the symbolic impoverishment and spiritual disenchantment of Alphaville is of course Lemmy Caution, the successor secret agent from the exterior who literally and figuratively, with his ever-present cigarette lighter, brings light to the Manichean darkness. But Godard's image of resistance is complicated by the fact that it is too obvious, too literal, to be taken seriously in any conventional sense. The figure of Lemmy Caution is over laden with too many evocative references to mid-twentieth-century popular culture: secret agents (e.g., James Bond), Dick Tracy, Guy l'Éclair (the French Flash Gordon), the legendary World War II battle of Guadalcanal, and finally the French cinematic legend of Lemmy Caution himself, a tough FBI agent, who had been played by Eddie Constantine in seven previous movies. More significantly, the tough-guy body language of Caution in his trench coat with upturned collar, ever-present cigarette, and ready gun conjures up a plethora of existentialist heroes and film noir protagonists. Indeed, Godard's Caution is such an explicit pastiche of film noir protagonists (there are noir shades of Sam Spade from John Huston's *The Maltese Falcon* [1941], Jeff Markham from Jacques Tourneur's *Out of the Past* [1947], and Mike Hammer from Robert Aldrich's *Kiss Me Deadly* [1955]) and an obvious self-parody of the earlier Lemmy Caution that one could easily conclude that he is a self-conscious camp hero who was conceived to deconstruct and kill off his more ingenuous camp predecessors.

Yet Caution represents more than a parody of what Sontag has called "the sensibility of failed seriousness" and "the theatricalization of experience" of camp, which defines much (but not all) of what is best in science fiction and film noir. As Fredric Jameson has pointed out, the concept of parody must be clearly distinguished from that of pastiche. Parody stands in tension and contrast with pastiche because parody depends upon the imitation, the mimicry, of a clearly established cultural norm or style in order to carry out its subversive intention. Pastiche, on the other hand, neither depends upon a well-defined norm nor expresses a subversive intention because it is an aesthetic form best suited to the postmodern era, which Sontag's early essays (especially those in *Against Interpretation* [1966]) and Godard's early films (especially *À bout de souffle* [1960]) helped to define. With the rise of a postmodern world, aesthetic heterogeneity has been succeeded by a more general proliferation of social codes and cultural fragmentation. "The norm itself is eclipsed." Consequently, as a postmodern form pastiche is a "neutral practice of . . . mimicry, without any of parody's ulterior motives, amputated of the satiric impulse, devoid of laughter and of any conviction that alongside the abnormal tongue you have momentarily borrowed, some healthy linguistic normality still exists."[16] In other words, one cannot parody popular art and naive camp in a postmodern age but only mimic them more or less well. From this perspective, Caution is a postmodern protagonist, a pastiche of pop styles, who illustrates "how to be a dandy in the age of mass culture"[17] but who nonetheless employs a camp sensibility to challenge the false seriousness and normality of the Alphaville regime.

As a camp protagonist, Caution does indeed draw upon the heritage of Baudelaire's ideal of the dandy, which runs throughout film noir. As the philosopher Stanley Cavell has argued, Baudelaire's ideal of the dandy has been an instructive presence in American film for some time: "Our most brilliant representatives of the type are the Western hero and Bogart; but we include the smaller and more jaded detectives and private eyes of the past generation: and the type is reiterated in the elegant nonprofessional solver of mysteries."[18] Like Baudelaire, Cavell identifies the feature of a "hidden" or "latent" fire as essential to the character of the dandy—what Baudelaire describes as "an air of coldness which comes from an unshakeable determination not to be moved . . . a latent fire which hints at itself, and which could, but chooses not to burst into flame."[19] But what Cavell fails to note is that the jaded protagonists of film noir are distinguished by their failure to keep their fires *banked*. The same apparent failure distinguishes Caution. But whereas in classic film noir the protagonist typically loses control of himself and his circumstances, Caution is depicted as having a free will. He ostensibly chooses to make the fire of his lighter manifest and to follow the promptings of his passion for Natasha Von Braun (Anna Karina).

Because he is a science fiction noir protagonist, Caution's rebellion against the darkness of Alphaville draws upon the tragic seriousness of high culture and especially the moral seriousness associated with "extreme states of feeling" to a significantly greater extent than does his successor Rick Deckard in Ridley Scott's *Blade Runner* (1982) in his quest for the criminal replicants and eventually the love of Rachael, even as both of these sensibilities of seriousness are deeply intertwined with a camp sensibility that predominates in the latter film. In contrast to that in *Blade Runner,* it is "the kind of seriousness whose trademark is cruelty, anguish, derangement," which is so prominent in French literary culture, that prevails in *Alphaville* (note Caution's reference to Céline's *Journey to the End of the Night*)—the quest for what André Breton succinctly characterized as "complete *nonconformism.*"[20] Specifically, the surrealist sensibility of Éluard's poetry saves Lemmy and Natasha from the prosaic nightmare of Paris sans art transmuted into Alphaville. As Adrian Martin has argued, the pivotal lyrical interlude in which the romance of Lemmy and Natasha is depicted in a complex set of scenes that rupture the film space-time continuum as they assume "dance-like postures," reciting a pastiche of Éluard's poetry, is a work of poetry in itself that illustrates the transformative power of art. "The lyrical transport provided by a poetic recital does not merely mirror the characters but directly transforms them: from a halting, uncomprehending delivery earlier in the scene, Natasha now magically moves to being a smooth, communicating vessel for verse (and Lemmy changes from a tough guy to a Bressonian model)."[21] Through poetry a new unity of image and language is achieved that permits the saving truth of love to be grasped. However, the exquisite *dérangement* of this love depends upon a poetic enactment that cannot be captured or expressed in words alone. For Godard, the transformative power of art is, not surprisingly, inseparable from the media of an electronic as opposed to print culture.

Dystopian Antivisions

Nonetheless, *Alphaville* clearly draws upon a well-known literary tradition of dystopian novels, best represented by Yevgeny Zamyatin's *We* (1924), Aldous Huxley's *Brave New World* (1932), and George Orwell's *1984* (1949). This literary tradition is notable for its success in depicting the political oppression and spiritual destructiveness of totalitarian societies and for its failure to articulate a compelling and ennobling vision of collective and individual life beyond these disenchanted worlds. For instance, Orwell's *1984* closes on a note of profound despair for the reader with an attitude of total acceptance toward the corruptions of political power. Winston Smith, after all, does not symbolically triumph over O'Brien but succumbs. In the end, Winston has moved beyond personal despair because he has been so completely emptied of memory and the capacity for love that there is nothing left to do but consummate his totalitarian surrender and "love" Big Brother. Winston's acquiescence to unchecked power imaginatively represents what Orwell elsewhere predicted in a mood of total despair: "The autonomous individual is going to be stamped out of existence. But this means that literature, in the form in which we know it, must suffer at least a temporary death."[22] Indeed, the death of literature and arts dependent upon a book culture is a standard theme in dystopian literature. But to imagine the death of creative cultural forms closely associated with a sophisticated print culture practically guarantees the emergence of the bleak and pessimistic perspective, the antivision, that seems endemic to the dystopian imagination, unless cultural forms are introduced that are nonliterary, such as the resort to oral culture in Ray Bradbury's novel *Fahrenheit 451* (1953) and the reliance on electronic culture in *Alphaville* to communicate a vision of how the world is to be endured and perhaps rebelled against.

The failures of print culture are evident everywhere in *Brave New World* and *1984*, but nowhere are they more evident than in the inarticulate and failed rebellions of those who challenge the totalitarian regimes. In Orwell's novel, Julia, like Natasha, also dares to assert "I love you." But Julia's rebellious assertion of love against the Party is sexual, not emotional, precisely because the Party suppresses sexuality in order to monopolize erotic identification with the Party. "Not love so much as eroticism was the enemy, inside marriage as well as outside it."[23] Consequently, Julia's insistence upon her own spontaneous sexuality is political. Engaging in sexual intercourse is a political act against the regime. In *Alphaville* and *Brave New World*, the reverse formulation holds: not unbridled eroticism but love is the enemy. In the latter, falling in love, indeed expressing any intense emotion, translates into a political act that risks punishment and execution. In fact, even crying is forbidden. In an ironic and perhaps deliberate reversal of the plight of Meursault in Camus' *The Stranger* (another outsider from the exterior), someone is executed *for* crying, in this case, at a spouse's funeral.

In other important aspects, Alphaville falls closer to the regime of Huxley's *Brave New World* insofar as various forms of sensual permissiveness, especially sexual promiscuity and drug use, are encouraged as a form of political control. Indeed, the very title of Godard's dystopia is apparently taken from the upper caste of Huxley's regime—the Alphas. Both dystopias resemble the regime of Dostoyevsky's Grand Inquisitor in their indulgence and encouragement of human shortcomings. In fact, both trade upon the sinister implication of the Grand Inquisitor as a director of sexual, even orgiastic, activities. "We shall allow or forbid them," says the Grand Inquisitor, "to live with their wives and mistresses, to have or not to have children—according to whether they have been obedient or disobedient—and they will submit to us gladly and cheerfully." In such a regime, politics and sexuality merge to form an image of eroticized power that sanctions "every sin."[24] Even so, both Huxley's Mustapha Mond and Orwell's O'Brien are descendants of the Grand Inquisitor, exercising erotic domination respectively in permissive and repressive fashions. Godard's conception of Alphaville draws upon this tradition of manipulation and coercion of the most private of activities.

Artistic and Erotic Salvation

Caution's rebellion in the name of the bridled eroticism of love succeeds where the unbridled eroticism of Julia in *1984* and the ascetic passion of the Savage, John (also from outside civilization), in *Brave New World* fail as modes of resistance against their respective regimes. Rather than suggesting that the violence of the passions requires ritual forms and rules to control and effectively express them, however, Godard implies that it is the liberation of the individual from repressive social forms that permits the spontaneous expressions of the spirit. Pitting the individual against such an obviously oppressive regime affirms the right and power of the individual. Creative sensibilities are linked with the surrealist quest for "complete nonconformism" rather than the forms of high culture, natural impulse rather than social conscience, even though Caution clearly embodies and appeals to the memories and reality of the past. Indeed, conscience itself is assimilated to the liberation of aesthetic expression and erotic desire.

Godard juxtaposes the sophisticated aesthetic and erotic expressiveness of Lemmy and Natasha to the dogma of scientific-technological progress represented by Alphaville. He suggests that the scientific-technical rationality of Alphaville gone awry can be defeated by what Weber calls the "this-worldly salvation" of art and eroticism, or more specifically an aestheticized eroticism. In Weber's seminal formulation concerning the advance of disenchantment, in fact, "art takes over the function of a this-worldly salvation. . . . It provides a *salvation* from the routines of everyday life, and especially from the increasing pressures of theoretical and practical rationalism." Likewise, in the erotic relation the lover knows himself or herself "to be freed from the cold skeleton hands of rational orders, just as completely as from the banality of everyday routine."[25] In *Alphaville,* Godard signals the beginning of this-worldly salvation in a "negative-printed sequence" near the end of the film—what Darke accurately characterizes as "a sign of transition from the land of the dead (Alphaville) to the land of the living (the Outerlands and beyond)."[26] After Lemmy poses the riddle that will eventually lead to the self-destruction of the Alpha 60 computer and commences his escape, black becomes white and white becomes black in the film, implying that he has begun to reverse the symbolic and spiritual darkness of this world with his daring rebellion and assertion of love.

But what Godard obscures, at least from the perspective of Weberian theory, is that the aesthetic and erotic salvation so brilliantly depicted in *Alphaville* is itself a manifestation and symptom of intellectualization and rationalization, a highly intellectualized response to the disenchantment of the world. With respect to the erotic sphere, Weber states, "As the knowing love of the mature man stands to the passionate enthusiasm of the youth, so stands the deadly earnestness of this eroticism of intellectualism to chivalrous love. In contrast to chivalrous love, this mature love of intellectualism reaffirms the natural quality of the sexual sphere, but it does so consciously, as an embodied creative power."[27] For Weber, the disenchantment of the world is by no means exhausted in the instrumental rationalism of science and technology; rather, the quest for inner-worldly salvation and perfection of self is deeply problematic and self-defeating. Like Godard, Weber shares an imagination of spiritual disaster. But unlike Weber, who insists that there is no escape from the disenchanted present, Godard leaves room for the hope of an enchanted future.

Notes

The first epigraph to this essay is taken from Susan Sontag, "Notes on 'Camp,'" in *Against Interpretation and Other Essays* (New York: Farrar, Straus and Giroux, 1966), 286–87.

The second epigraph comes from Paul Éluard, "Poetic Evidence," in *Surrealism*, ed. Herbert Read (London: Faber and Faber, 1936), 172.

1. Susan Sontag, "The Imagination of Disaster," in *Against Interpretation*, 212–13.

2. See Allen Thiher, "Postmodern Dilemmas: Godard's *Alphaville* and *Two or Three Things That I Know about Her*," *Boundary 2* 4 (1976): 947–56.

3. See Marc Augé, *Introduction to an Anthropology of Supermodernity*, trans. John Howe (London: Verso, 1995), 75–115.

4. Chris Darke, *Alphaville* (Urbana and Chicago: University of Illinois Press, 2005), 30, 33.

5. Peter Wollen, *Paris Hollywood: Writings on Film* (London: Verso, 2002), 201–4.

6. Vivian Sobchack, "Lounge Time: Postwar Crises and the Chronotope of Film Noir," in *Refiguring American Film Genres: Theory and Method*, ed. Nick Browne (Berkeley: University of California Press, 1998), 129, 148.

7. M. M. Bakhtin, *The Dialogic Imagination: Four Essays*, trans. Caryl Emerson and Michael Holquist, ed. Michael Holquist (Austin: University of Texas Press, 1981), 84.

8. R. Barton Palmer, "'Lounge Time' Reconsidered: Spatial Discontinuity and Temporal Contingency in *Out of the Past*," in *Film Noir Reader 4*, ed. Alain Silver and James Ursini (New York: Limelight, 2004), 57–58.

9. Ibid., 58.

10. Henri Bergson, *Time and Free Will: An Essay on the Immediate Data of Consciousness*, trans. F. L. Pogson (New York: Humanities Press, 1971), 100.

11. Max Horkheimer, letter, October 14, 1942, quoted in Martin Jay, *The Dialectical Imagination: A History of the Frankfurt School and the Institute of Social Research, 1923–1950* (Boston: Little, Brown, 1973), 214 (emphasis added).

12. See Michel Marie, *The French New Wave: An Artistic School*, trans. Richard Neupert (Malden, MA: Blackwell, 2003), and Richard Neupert, *A History of the French New Wave Cinema* (Madison: University of Wisconsin Press, 2002).

13. Max Weber, "Science as a Vocation," in *From Max Weber: Essays in Sociology*, trans. and ed. H. H. Gerth and C. Wright Mills (New York: Oxford University Press, 1958), 138–40.

14. Charles Taylor, *Sources of the Self: The Making of Modern Identity* (Cambridge, MA: Harvard University Press, 1989), 4, 19, 17. This paragraph and the following one are adapted from my essay "The Horizon of Disenchantment: Film Noir, Camus, and the Vicissitudes of Descent," in *The Philosophy of Film Noir*, ed. Mark T. Conard (Lexington: University Press of Kentucky, 2006), 109–10.

15. Taylor, *Sources*, 17–18, 94.

16. Fredric Jameson, *Postmodernism, or The Cultural Logic of Late Capitalism* (Durham, NC: Duke University Press, 1991), 17.

17. Sontag, "Notes on 'Camp,'" 288.

18. Stanley Cavell, *The World Viewed: Reflections on the Ontology of Film,* enlarged ed. (Cambridge, MA: Harvard University Press, 1979), 56.

19. Charles Baudelaire, *The Painter of Modern Life and Other Essays,* trans. and ed. J. Mayne (New York: Da Capo, 1964), 29.

20. André Breton, "Manifesto of Surrealism," in *Manifestoes of Surrealism,* trans. Richard Seaver and Helen R. Lane (Ann Arbor: University of Michigan Press, 1972), 47.

21. Adrian Martin, "Recital: Three Lyrical Interludes in Godard," in *Forever Godard,* ed. Michael Temple, James S. Williams, and Michael Witt (London: Black Dog, 2004), 263. Robert Bresson, both a predecessor and contemporary of the young French new wave directors, was noted for his rejection of conventional cinema and psychological motives and greatly admired for his uncompromising filmmaking. See Marie, *French New Wave,* and Neupert, *History of the French New Wave Cinema.*

22. George Orwell, "Inside the Whale," in *Inside the Whale and Other Essays* (New York: Penguin Books, 1976), 48.

23. George Orwell, *1984* (New York: Signet Classic, 1961), 57.

24. Fyodor Dostoyevsky, *The Brothers Karamazov,* trans. Constance Garnett (New York: Signet Classic, 1957), 239.

25. Max Weber, "Religious Rejections of the World and Their Directions," in *From Max Weber,* 342, 347.

26. Darke, *Alphaville,* 59.

27. Weber, "Religious Rejections of the World," 347.

THE MATRIX, THE CAVE, AND THE COGITO

Mark T. Conard

Thomas Anderson, computer programmer and hacker, learns that everything he thought he knew about the world and his life is false, that he's been deceived. Further, he discovers that he and most of his fellow human beings are enslaved in a way that he never could have imagined and that he is the chosen One, the savior who will lead them out of the slavery of ignorance and to enlightenment and understanding. Interestingly, René Descartes asks us to imagine a similar all-encompassing deception, and Plato famously writes in the *Republic* about just such an escape from bondage and a journey to enlightenment.

Indeed, *The Matrix* (Andy Wachowski and Larry Wachowski, 1999) is a smart film. With its clearly intended references, not only to Descartes and Plato but also to the Bible and Buddhism, to Lewis Carroll, to Baudrillard, and to Orwell, it may in fact be a deep film. But how deep does it go? The film looks as if it has a metaphysics and an epistemology of its own that are akin to Plato's and Descartes'. I argue below that because the film rejects any notion of transcendence, it only references or borrows from Plato and Descartes, without staying true to their spirit; I argue that it doesn't go deep enough.

First, and briefly, what are metaphysics and epistemology? These are branches or subdisciplines within philosophy. Metaphysics concerns the nature of reality; questions about whether God exists, whether we have free will, and whether the mind is a different substance from the body—or, for that matter, whether it is a substance at all—are metaphysical concerns. On the other hand, and probably less familiarly, epistemology is the study of the nature, sources, and grounds of knowledge, and so issues about the nature of truth and falsity and how to distinguish knowledge from opinion are epistemological matters. It's important to note that one's metaphysics and epistemology go hand in hand: you can't talk about the nature of reality without at the same time making knowledge and truth claims, and when you do make knowledge and truth claims, those claims concern some feature of the world; they're about some bit of reality and thus presuppose a metaphysics.

Before I begin, I must make a few admissions or caveats. First, my discussion here will be restricted to the first film, *The Matrix*, and will not concern the sequels, *The Matrix Reloaded* and *The Matrix Revolutions* (both Andy Wachowski and Larry Wachowski, 2003). I do this because, on the one hand, I believe the first film stands on its own as a complete, coherent story, and, on the other hand, the sequels at best add nothing of value to the narrative and at worst contain ridiculous and absurd plot developments and resolutions and thus undermine the brilliance and originality of the first film when they're all packaged together.[1] The rumor seems to be that the Wachowski brothers had a trilogy in mind and sketched out from the beginning. There are those of us who are skeptical of this claim, given the length of time it took for them to release the other two films after the first one was released, and given the sublimity of the original film in comparison to the silliness of the sequels. My second admission is that I realize that some of what I'm talking about here—Platonic, Cartesian, and Christian elements or influences in the film—is very well-trodden ground.[2] However, on the one hand, I don't think there's any such thing as too much discussion of Plato or Descartes, and, on the other hand, the conclusions I'm going to draw about the metaphysics and epistemology of the film are at odds with other literature on the subject.

What Is the Matrix?

Before I discuss Plato and Descartes, I want to give a rundown of the plot of *The Matrix* for those unfortunate souls who might not have seen it. Thomas Anderson (Keanu Reeves), whose hacker alias is Neo, is contacted by supposed cyberterrorists Trinity (Carrie-Anne Moss) and Morpheus (Laurence Fishburne), who tell him that his suspicions that there's something out of joint with the world are well founded and that he's in danger. Morpheus hints vaguely at some important role Neo is to play and asks him if he really wants to know what the matrix is. Neo says he does.

Morpheus helps Neo awaken from the dream that has been his life to find himself nestled in a jelly-filled pod, in a massive field containing countless other pods. He has cables attached to his body, draining it of energy. He is delivered birthlike from his pod and taken aboard a hovercraft to find Morpheus, Trinity, and their colleagues on the ship.

Morpheus informs Neo that the age isn't the late twentieth century, as Neo had thought; it's as much as a hundred years later. In the early twenty-first century, computers developed artificial intelligence to the point of becoming autonomous. There ensued a war between humans and the computers, which the computers won. Ever after, the computers have been growing human beings as energy sources for themselves. The matrix, then, is a super-sophisticated computer program into which most of humanity is plugged, to keep them docile and ignorant while their energy is being harvested. The rebels, those who have been freed from the matrix and their ignorance, live aboard hovercrafts and can reinsert themselves into the matrix as they wish (their physical bodies of course remain aboard the ships; connecting cables are plugged into their brains, allowing them to enter the computer program). Within the matrix there are very deadly "sentient programs" known as agents, designed to hunt down rebels. No human, Neo is informed, has ever survived a direct confrontation with an agent.

Morpheus tells Neo that when the matrix was built, there was a man who had the power to change the program, to remake it as he saw fit, and it was he who freed the first of the rebels. After he died, the Oracle (Gloria Foster) prophesied his return, and ever after the rebels have been searching for his reincarnation, the One. Morpheus believes that Neo is in fact the One, the savior.

After much training in how to operate in the matrix and how to fight and use weapons, Neo is taken to see the Oracle, who tells him that a situation will present itself in which he will have to choose between saving Morpheus's life and saving his own.

Morpheus is captured by agents, and in rescuing him Neo discovers that he can do certain surprising things in the matrix, like dodge bullets. Having saved Morpheus and seen him safely returned to the hovercraft, Neo is pursued by agents, one of whom shoots him, apparently to death. Aboard the hovercraft, Trinity watches in horror and then declares her love for Neo and kisses him. He miraculously comes back to life to find that he can now control the matrix and destroy agents at will. Reborn with these new powers, he apparently confirms Morpheus's belief that he, Neo, is the One.

Let us now turn to the influence of Plato and Descartes on *The Matrix*.

Plato's Forms

One major metaphysical concern of ancient Greek philosophers was the nature of change in the world. They struggled to understand why and how the physical world around us seems to be perpetually in motion, perpetually changing, while at the same time certain elements of reality never (or seem never to) change. This problem or question is often framed in terms of what are called universals and particulars. Particulars are the everyday stuff we experience in the world around us—desks, tables, cars, etc.; they're physical, material things you can touch, see, hear, and taste. Universals, on the other hand, are the class categories into which these things fall—Desk, Table, Car, etc.—and are abstract rather than concrete. So, and coming back to the issue of change, while individual particular desks come into existence (they're manufactured), exist for a while, ultimately deteriorate like any physical thing, and then are destroyed, the universal Desk (or Deskness) seems to remain the same throughout, since the nature of a desk doesn't change when a particular desk or set of desks no longer exists.

Plato's theory of Forms, then, is the core of his metaphysics and it is a theory of universals. Note that individual, particular things are sensible—we perceive them with our five senses, but universals (class concepts, the commonality between things) are intelligible; we grasp them with our minds. In his *Phaedo*, Plato has Socrates and another character, Simmias, discuss the nature of the Forms:[3]

> Socrates: What about the following, Simmias? Do we say that there is
>   such a thing as the Just itself, or not?
> Simmias: We do say so, by Zeus.
> Socrates: And the Beautiful, and the Good?
> Simmias: Of course.
> Socrates: And have you ever seen any of these things with your eyes?
> Simmias: In no way . . .
> Socrates: Or have you ever grasped them with any of your bodily
>   senses? I am speaking of all things such as Bigness, Health,

Strength, and, in a word, the reality of all other things, that which each of them essentially is. Is what is most true in them contemplated through the body, or is this the position: Whoever of us prepares himself best and most accurately to grasp that thing itself which he is investigating will come closest to the knowledge of it?

Simmias: Obviously.

Socrates: Then he will do this most perfectly who approaches the object with thought alone, without associating any sight with this thought, or dragging in any sense perception with his reasoning, but who, using pure thought alone, tries to track down each reality pure and by itself, freeing himself as far as possible from eyes and ears and, in a word, from the whole body, because the body confuses the soul and does not allow it to acquire truth and wisdom whenever it is associated with it.[4]

Plato doesn't typically talk about physical objects like desks and chairs. When discussing the Forms, he usually mentions the virtues (Piety or Justice, for example) and, as he does here, Beauty and Goodness (and somewhat less often does he refer to concepts like Bigness and Health). The point he is making in this passage, as I mentioned above, is that we don't grasp the Forms with our bodily senses but rather with our minds, and, he argues, we grasp them best and most purely the less the body has to do with the whole affair.

These Forms are the unchanging, eternal essences of particulars. His language is (probably necessarily) vague, but Plato says that particulars "participate" in the Forms. Particulars have the qualities they have, beauty, goodness, strength, etc., because they partake of the Forms. It's the Forms that make particulars what they are. The second premise to this argument is that what's ultimately real is what endures, what is lasting. Consequently, since particulars are ephemeral, doomed to pass away, and the Forms are eternal, the latter are ultimate reality. Let's be clear: Many of us believe that what's real is what's physical, what you can get your hands on, what you can see, etc., perhaps to the point of thinking of classes or categories (species, genera, etc.) as mere mental constructions, useful abstractions; Plato, on the contrary, thinks this is exactly wrong. What you can touch, taste, or smell is, for that very reason, less real. It's what you grasp with your intellect, removed from the body, that's ultimately real.

The Cave

Plato's famous cave allegory in book 7 of the *Republic* is a depiction of the intellectual journey one takes from perceiving particulars to grasping the Forms intellectually. In the passage, Socrates describes prisoners living in a cave:

> Socrates: Imagine human beings living in an underground, cavelike dwelling. . . . They've been there since childhood, fixed in the same place, with their necks and legs fettered, able to see only in front of them, because their bonds prevent them from turning their heads around. Light is provided by a fire burning far above and behind them. Also behind them, but on higher ground, there is a path stretching between them and the fire. Imagine that along this path a low wall has been built, like the screen in front of puppeteers above which they show their puppets.
>
> Glaucon: I'm imagining it.
>
> Socrates: Then also imagine that there are people along the wall, carrying all kinds of artifacts that project above it—statues of people and other animals, made out of stone, wood, and every material. And, as you'd expect, some of the carriers are talking, and some are silent.
>
> Glaucon: It's a strange image you're describing, and strange prisoners.
>
> Socrates: They're like us. Do you suppose, first of all, that these prisoners see anything of themselves and one another besides the shadows that the fire casts on the wall in front of them?
>
> Glaucon: How could they, if they have to keep their heads motionless throughout life? . . .
>
> Socrates: And what if their prison also had an echo from the wall facing them? Don't you think they'd believe that the shadows passing in front of them were talking whenever one of the carriers passing along the wall was doing so?
>
> Glaucon: I certainly do.
>
> Socrates: Then the prisoners would in every way believe that the truth is nothing other than the shadows of those artifacts.[5]

Because the prisoners have only ever had experience of the shadows on the cave wall—they can see nothing else—they believe that those shadows constitute the whole of reality. We're like the prisoners in the cave when we believe that the physical, material world of particulars is the whole of reality, and that there's nothing else beyond. The journey out of the cave for us, what may be called enlightenment, consists in leaving the body and senses behind and making the intellectual ascent to the Forms.

Neo's Journey out of the Cave

This intellectual journey is not an easy one. If one were to release a prisoner, says Socrates, and show him the fire or the artifacts, something that was real, not only would he be incredulous that he was experiencing reality, he would be angry and want to return to his comfortable shadow world:

> Socrates: Consider, then, what being released from their bonds and cured of their ignorance would naturally be like if something like this came to pass. When one of them was freed and suddenly compelled to stand up, turn his head, walk, and look up toward the light, he'd be pained and dazzled and unable to see the things whose shadows he'd seen before . . .
> And if someone compelled him to look at the light itself, wouldn't his eyes hurt, and wouldn't he turn around and flee towards the things he's able to see, believing that they're really clearer than the ones he's being shown?
> Glaucon: He would.
> Socrates: And if someone dragged him away from there by force, up the rough, steep path, and didn't let him go until he had dragged him into the sunlight, wouldn't he be pained and irritated at being treated that way? And when he came into the light, with the sun filling his eyes, wouldn't he be unable to see a single one of the things now said to be true?[6]

The prisoners are so thoroughly trapped in their ignorance, so convinced of the reality of the shadows, that they'd have a very difficult time believing their liberators; they would in fact at first resent being liberated and want nothing more than to return to their ignorance.

The Wachowskis' use of, or reference to, Plato's cave allegory in *The Matrix* is clear. The prisoners in the cave are represented in the film of course as the hapless humans plugged into the matrix, nestled in their pods, providing energy for the computers. Just as the prisoners believe that the shadows are real, so too we in the matrix believe that the computer-generated illusion we experience as our daily lives is reality. When Neo is unplugged and taken out of the matrix to face "the desert of the real," like Socrates' prisoners, he's "pained and dazzled" and asks, "Why do my eyes hurt?" and Morpheus, very Socratically, answers, "Because you've never used them before." Again, to leave the matrix is to make the ascent out of the cave; it is to leave the shadows behind and experience reality. Neo has made that journey, and he has some moments of incredulity at first, has a hard time believing what Morpheus is showing him, just as Socrates claims the prisoners in the cave would.[7]

Descartes' Doubting Method

In addition to being a revolutionary philosopher, René Descartes was a great mathematician, and this is reflected at times in his philosophy. For example, in one of his central works, *Meditations on First Philosophy,* he wants to construct a metaphysics on the model of a mathematical system. Starting from some indubitable axiom or axioms (as we do in geometry, say), he'll deduce everything from these truths, and if his beginning axioms are in fact undoubtedly true, and if his proofs are rigorous and valid, then his conclusions have to be true.

This is reasonable enough, but how does one find, come up with, or generate an indubitable axiom—some truth that is beyond all question? And we're not talking about some garden variety of certainty here; Descartes wants metaphysical, absolute certainty, logical certainty. I know, for example, that I locked the door of my apartment this morning. I'm certain of this. But is this *logical* certainty? Isn't it within the realm of possibility that I could be mistaken, and that I left the door unlocked? Well, sure it is. So this is merely contingent knowledge, it's nonnecessary, and this isn't the kind of certainty we need. We need, says Descartes, a certainty on the level of mathematics, the $2+2=4$ kind of certainty. We want a priori knowledge, knowledge that's universal and necessary.

In order to find this axiom, this one undoubtable truth upon which he will found his metaphysics, then, Descartes famously employs the doubting method. He will doubt everything to find the one thing that can't possibly be doubted. He begins by doubting his sense experience—what he hears, sees, feels, etc.—and wonders if he could be asleep and dreaming all that he takes to be real. He (initially) takes this to be possible and concludes that "there are no definitive signs by which to distinguish being awake from being asleep."[8] If this were so, then a great deal of what we think we know about the world and about ourselves would be thrown into doubt.

Interestingly, when discussing with Neo for the first time the nature of reality, Morpheus (in Greek mythology the god of dreams), nearly quoting Descartes, asks, "Have you ever had a dream, Neo, that you were so sure was real? What if you were unable to wake from that dream? How would you know the difference between the dream world and the real world?" And, as important, if you can't wake from the dream or distinguish it from reality, then almost everything you know about the world and human existence, or thought you knew, could turn out to be false.

The Evil Genius

The doubt generated by the dream thesis isn't quite radical enough for Descartes. In the end, he pushes the doubting method to its extreme limit by imagining an evil, godlike genius deceiving him at every instant of his life: "Accordingly, I will suppose not a supremely good God, the source of truth, but rather an evil genius, supremely powerful and clever, who has directed his entire effort at deceiving me. I will regard the heavens, the air, the earth, colors, shapes, sounds, and all external things as nothing but the bedeviling hoaxes of my dreams, with which he lays snares for my credulity."[9] With an all-powerful, but evil, god implanting perceptions, thoughts, and memories directly into my mind, not just my whole life but the entire world around me could be a complete fiction.

The references to Descartes are perhaps more subtle than the ones to Plato in *The Matrix*, but the influence is there. The "singular consciousness" that generates the matrix is the evil genius, and nearly all of humanity in the film finds itself in the unfortunate position of being subject to the "bedeviling hoaxes" of that consciousness. So, to return to Descartes' point, would it be logically possible for everything we experience around us to be part of a sophisticated computer program, which is input directly into our brains, while we slumber peacefully in jelly-filled cocoons, generating power to run the computers? If so, then, again, the vast majority of what we think we know about the world could be false.

Descartes' Indubitable Principle and the Ascent out of the Cave

While clearly *The Matrix* contains these allusions to Plato and Descartes, the question I want to address is whether the film has a Platonic or Cartesian metaphysics and epistemology, whether it stays true to the spirit of Plato and Descartes. As I mentioned above, I don't think it does, and to see why let's think about both the conclusion to Descartes' doubting method and then the meaning behind Plato's cave metaphor.

First, as I said, with the evil genius deceiving me, everything I know about the world around me and all my thoughts, feelings, perceptions, and memories might be false. It might be that I don't even have a body, for instance, since I'm aware of my own body through sense perception. So, in pushing his doubt to its most extreme limit, Descartes poses the ultimate skeptical question: is it also possible that he, Descartes, does not exist? He answers, "Then too there is no doubt that I exist, if he is deceiving me. And let him do his best at deception, he will never bring it about that I am nothing so long as I shall think that I am something. Thus, after everything has been most carefully weighed, it must finally be established that this pronouncement 'I am, I exist' is necessarily true every time I utter it or conceive it in my mind."[10] Descartes' famous phrase is "Cogito, ergo sum" (I think, therefore I am), which he uses elsewhere, not in the *Meditations.* The point, and the answer to his question, is that the evil genius could make him doubt his own existence but that he must exist in order to do the doubting. He thinks, that is, he has an intellectual intuition—in Descartes' language, a clear and distinct idea—of himself as thinking, and therefore he knows that he must exist. If there's thinking going on, there must be a thinker. This is his one absolutely indubitable assertion, the one thing he knows with metaphysical certainty: that he exists as a mind, as a thinking thing.[11]

Descartes later affirms in the *Meditations* that he similarly has an intellectual intuition of God: "Of all the ideas that are in me, the idea that I have of God is the most true, the most clear and distinct." And in fact he goes on to claim, perhaps surprisingly, that the idea of God is so immediate that it precedes even his perception of himself: "Thus the perception of the infinite is somehow prior in me to the perception of the finite, that is, my perception of God is prior to my perception of myself."[12] Descartes doesn't here mean the intuition of God is temporally prior, that he somehow perceives God before he's aware of himself, but rather that it is (somehow) logically prior, that the awareness of the finite is possible only against the background, so to speak, of the infinite.[13]

Ultimately, then, it is the immediate intellectual apprehension of God as a clear and distinct idea that guarantees the truth of all of Descartes' other ideas and perceptions. He knows that he's not being deceived about ordinary perceptions, and he knows that his understanding of the universe is accurate, because God is all-powerful and all-good, God created him and his intellectual faculties, and God would not deceive him.

In the cave allegory Plato is similarly describing for us how to attain necessary knowledge. As I mentioned earlier, the ascent out of the cave is a metaphor for the mind's movement away from contingent (nonnecessary) sensory information (everyday truths about the changing world that could easily be falsified) to a pure, intellectual intuition of the Forms. In other words, what we can grasp and understand about the physical, material world (the shadows on the cave wall) is mere contingent knowledge. We have real—universal and necessary—knowledge only of the eternal and unchanging universal Forms. Leaving the cave and seeing reality for the first time, then, is a metaphor for leaving behind the body, the senses, and our perceptions and attaining a pure, intellectual apprehension of the Forms and thus gaining a priori knowledge, knowledge that's certain and that is beyond empirical verification, like those mathematical truths I mentioned earlier. You don't need to go around and check every time you put two apples together with two more apples to make sure that you come up with four apples. This is something that you know immediately and with absolute certainty. It doesn't require empirical verification.

I argue below that, while *The Matrix* has its allusions to Descartes and Plato, it abandons any notion of transcendence and any concept of the a priori and is thus contrary to the spirit of Cartesian or Platonic metaphysics and epistemology.

The Matrix **as the Christ Story**

As opposed to relying on faith, Descartes argues that he has an intellectual intuition, a rational grasping, of God, which assures him that his perceptions are correct, that he's not being deceived by the evil genius, and that he can thus have a priori, universal and necessary, knowledge about the world, the certainty of which is grounded in the nonempirical. *The Matrix*, as we'll see, has no such recourse.

First, let's note that the basic outline of the film's narrative is the Christ story: a man is chosen as savior, dies, is resurrected, and then is able to save humanity (or so we suspect by the end of the film, when he's able to destroy the agents and fly away like Superman). But, while the film has its Christian trappings—it makes use of Christian symbols and iconography—it has undermined the essential meaning of those symbols by abandoning the deity. That is, in presenting Neo as the Christ, *The Matrix* is Christianity without God.[14] Neo is indeed the One; he dies and is resurrected. But what being the One amounts to is being able to "remake the matrix" as he sees fit; he can bend and break the rules of the program, for reasons that go unexplained. But the latter is just a computer-generated dream world, the unreal, and in reality—in the real world outside the matrix, in the desert of the real—Neo is just another man.[15] This would be equivalent, in Plato's terms, to saying he's the One because he can make the shadows appear when he wishes, which is perhaps a powerful and useful ability within the cave, but of course it at no point touches ultimate reality or the transcendent. On the other hand, Morpheus is the Father (or at least is so described by Tank, another crew member aboard the *Nebuchadnezzar*, Morpheus's ship), but he too is just another man, and he doesn't even have the powers of the One in the matrix. Last, Trinity apparently completes the triad, but rather than being the Holy Ghost, she's more like Mary Magdalene. Again, the film contains the story and symbolism of Christianity without the transcendent, without God. Further, the film seems to reject with God any other notion of transcendence as well, and thus any foundation for a priori knowledge, and because of that its use of Descartes and Plato belies the true spirit of those two thinkers.

On the contrary, for the denizens of the matrix there is no intellectual intuition of God as proof of the possibility of universal and necessary knowledge; the only proof of reality they're offered is what one can see, hear, smell, taste, experience for oneself. When experiencing the "loading program" for the first time, for example, Neo asks, "This isn't real?" And Morpheus responds, "What *is* real? How do you define 'real'? If you're talking about what you can feel, what you can smell, what you can taste and see, then 'real' is simply electrical signals interpreted by your brain." Morpheus seems to be suggesting here that reality is circumscribed and perhaps defined by what one can directly sense and perceive, that there's nothing beyond what we can tap into with our five senses. And this is what I mean by a rejection of any notion of transcendence.

But note that, interestingly enough, this is precisely the kind of empiricism that generated the problem that Descartes was facing and trying to overcome in the first place. Not only are the senses not the source of our most important knowledge, the a priori, they're not even reliable much of the time. And that's exactly what led Descartes to the doubting method and to his search for the one thing that can't be doubted.

Consequently, we might ask, What's the proof that what Morpheus describes as reality, life outside the matrix, life aboard the hovercraft, isn't simply another computer-generated fiction to keep us docile? Or what's the proof that it isn't some elaborate dream that Neo's having while nestled quietly in his pod?[16]

The only answer offered to these questions would seem to be that one "just knows." Upon taking Neo out of the matrix, for example, Morpheus says, "We've done it, Trinity. We've found him." Trinity responds, "I hope you're right." And Morpheus tells her, "I don't have to hope. I know it." But upon what is his knowledge based? How does he "know"? He offers no evidence, no proof, no argument that Neo is the One. He simply seems to feel it is so. Indeed, this is the Oracle's answer to how Neo will "know" if he's truly the One. She tells him, "I'm going to let you in on a little secret. Being the one is just like being in love. No one can tell you you're in love, you just know it, through and through, balls to bones."

Thus knowledge and certainty in *The Matrix* are based on a kind of intuition that one has about things. But this isn't an intellectual intuition, a rational grasping of the a priori, and therefore something objective. Rather, it's a gut feeling, it's like being in love, it's completely subjective. And of course subjective states are proof enough of our own feelings, thoughts, or conditions but don't necessarily say anything at all about reality.[17]

Note, then, that this emphasis on perceptions and feelings, this fetishizing of the empirical, was precisely what Plato was describing as living in the cave. Plato was telling us that what we sense and feel isn't proof of anything, and it isn't the source of our most important knowledge, the a priori. In taking his feelings as absolute proof, Neo is like a prisoner in Plato's cave waking from a dream, seeing the shadows on the wall, and taking that as proof that he's no longer dreaming.

Notes

Many thanks to Steven M. Sanders for his very helpful comments and suggestions on earlier drafts of this essay.

1. The idea, for example, implied in the third film, that Neo is literally divine and can change the world outside the matrix is simply insulting.

2. See, for example, the excellent The Matrix *and Philosophy: Welcome to the Desert of the Real,* ed. William Irwin (Chicago: Open Court, 2002).

3. Plato writes dialogues, narrative pieces in which the main character is often Socrates. So, while it's Socrates speaking here, Plato is the author of the words. In all the quoted passages from Plato's dialogues, for clarification I have added the speakers' names with a colon (e.g., "Socrates:"). These are not found in the original text or in the translations from which I am quoting.

4. Plato, *Phaedo,* in *Five Dialogues,* trans. G. M. A. Grube (Indianapolis, IN: Hackett, 2002), 65d–e.

5. Plato, *Republic,* trans. G. M. A. Grube (Indianapolis, IN: Hackett, 1992), bk. 7, 514a–515c.

6. Ibid., 515c–516a.

7. Note, too, that we're meant to think of the Oracle as the oracle at Delphi, who announced that Socrates was the wisest man alive, as recounted in Plato's *Apology.* The dictum, "Know thyself," hanging on the kitchen wall of the Oracle was indeed inscribed on the temple at Delphi. This is another obvious reference to the ancient Greeks and an indirect reference to Socrates and Plato.

8. René Descartes, "Meditation I: Of the Things of Which We May Doubt," in *Meditations on First Philosophy,* in *Modern Philosophy: An Anthology of Primary Sources,* ed. Roger Ariew and Eric Watkins (Indianapolis, IN: Hackett, 1998), 28.

9. Ibid., 29.

10. Descartes, "Meditation II: Of the Nature of the Human Mind," in *Meditations,* 30.

11. That Descartes is able to doubt the existence of his body but not his mind confirms his belief that mind and body are radically different substances. This belief is known as substance dualism, the idea that the world (and we ourselves) is made up of two different kinds of stuff. Substance dualism generates the unfortunate mind/body problem: the question of how mind and body interact if they're such radically different sorts of things.

12. Descartes, "Meditation III: Of God," in *Meditations,* 38.

13. "For how would I understand that I doubt and that I desire, that is, that I lack something and that I am not wholly perfect, unless there were some idea in me of a more perfect being, by comparison with which I might recognize my defects." Ibid.

14. Gregory Bassham says, "Although *The Matrix* contains many obvious Christian motifs, it is by no means a 'Christian movie.' Rather, it is a syncretistic tapestry of themes drawn from Tibetan and Zen Buddhism, Gnosticism, classical and contemporary Western epistemology, pop Quantum mechanics, Jungian psychology, postmodernism, science fiction, Hong Kong martial arts movies, and other sources.

"The film features a decidedly non-Christian conception of the Messiah. According to orthodox Christian belief, Jesus was a sinless God-man who brought salvation to the world, not through violence or power, but through his sacrificial death and resurrection. Neo, by contrast, is a mere human being; he is far from sinless; he employs violence to achieve his ends (including, arguably, the needless killing of the innocent); and although he may bring liberation from physical slavery and mental illusion, he does not bring true salvation." Gregory Bassham, "The Religion of *The Matrix* and the Problems of Pluralism," in Irwin, The Matrix *and Philosophy,* 114.

15. As I said earlier, if the second and third films undermine my argument here, then so be it.

16. To my mind, this would have been the only satisfactory conclusion of the sequels: that, rather than being literally divine, Neo came to realize that life outside the matrix was itself another computer-generated dream world designed to keep the rebels complacent, and that he led them to escape for the first time to reality.

17. This is at odds with William Irwin's reading of the film: "Morpheus tells Neo that no one can be told what the Matrix is. You have to 'see it for yourself.' As with the Forms, it is not a literal 'seeing' but a direct knowing that brings understanding of the Matrix. . . .

"Neo too learns that intellect is more important than the senses. Mind is more important than matter. As for Plato the physical is not as real as the Form, so for Neo 'there is no spoon.'" William Irwin, "Computers, Caves, and Oracles: Neo and Socrates," in Irwin, The Matrix *and Philosophy,* 11.

CONTRIBUTORS

JEROLD J. ABRAMS is assistant professor of philosophy at Creighton University in Omaha, Nebraska. His essays appear in the journals the *Modern Schoolman*, *Philosophy Today*, *Human Studies*, and the *Transactions of the Charles S. Peirce Society* and in the volumes *James Bond and Philosophy* (Open Court, 2006), *Woody Allen and Philosophy* (Open Court, 2004), *Star Wars and Philosophy* (Open Court, 2005), and *The Philosophy of Film Noir* (University Press of Kentucky, 2006).

SHAI BIDERMAN is a doctoral candidate in philosophy at Boston University, a visiting fellow at the Institut für die Wissenschaften vom Menschen (Institute for Human Sciences) in Vienna, Austria, and an instructor in the College of Management in Israel. His research interests include philosophy of culture, philosophy of film and literature, aesthetics, ethics, existentialism, and Nietzsche. His essays appear in *Movies and the Meaning of Life* (Open Court, 2005), *Hitchcock and Philosophy* (Open Court, 2007), *South Park and Philosophy* (Blackwell, 2007), and in the forthcoming *The Philosophy of TV Noir* (University Press of Kentucky), *Star Trek and Philosophy* (Open Court), *Family Guy and Philosophy* (Blackwell), and *Lost and Philosophy* (Blackwell).

MARK T. CONARD is assistant professor of philosophy at Marymount Manhattan College in New York City. He's the coeditor of The Simpsons *and Philosophy* (Open Court, 2001) and *Woody Allen and Philosophy* (Open Court, 2004) and the editor of *The Philosophy of Film Noir* (2006), *The Philosophy of Neo-Noir* (2007), and *The Philosophy of Martin Scorsese* (2007), all published by the University Press of Kentucky. He's the author of "*Kill Bill: Volume 1*, Violence as Therapy," "*Kill Bill: Volume 2*, Mommy Kills Daddy," and "*Pulp Fiction:* The Sign of the Empty Symbol," all published on Metaphilm.com. He is also the author of the novel *Dark as Night* (Uglytown, 2004).

WILLIAM J. DEVLIN is assistant professor of philosophy at Bridgewater State College in Massachusetts. His fields of interest are philosophy of science, Nietzsche, existentialism, Eastern philosophy, and aesthetics. He has written on these areas in relation to such films and shows as *Star Trek, Lost, Family Guy, South Park,* and *The Prisoner.*

Jason Holt is assistant professor of communications at Acadia University in Nova Scotia. He is the author of *Blindsight and the Nature of Consciousness* (Broadview Press, 2003), which was short-listed for the 2005 Canadian Philosophical Association Book Prize, two novels, four books of poetry, and a number of articles, both scholarly and popular.

Deborah Knight is associate professor of philosophy at Queen's University in Ontario. Her main research area is philosophy of art, with emphases on literature and film. Recent publications include chapters in *Philosophy and Film and Motion Pictures* (Blackwell, 2006), the *Oxford Handbook of Aesthetics* (Oxford University Press, 2003), *Dark Thoughts: Philosophical Reflections on Cinematic Horror* (Scarecrow, 2003), and *Literary Philosophers? Borges, Calvino, Eco* (Routledge, 2002).

George McKnight is associate professor in the film studies program of the School for Studies in Art and Culture at Carleton University in Ontario. He edited *Agent of Challenge and Defiance: The Films of Ken Loach* (Greenwood, 1997) and has published articles on British cinema. With Deborah Knight, he has coauthored papers on *American Psycho*, *The Matrix*, Hitchcock's use of suspense, and detective narratives.

Jennifer L. McMahon is associate professor of English and philosophy and chair of the Department of English and Languages at East Central University in Oklahoma. Her areas of specialization include existentialism, philosophy and literature, aesthetics, non-Western philosophy, and biomedical ethics. She has published articles in journals including *Asian Philosophy* and the *Journal of the Association for Interdisciplinary Study of the Arts*. She has also published essays on philosophy and popular culture in Seinfeld *and Philosophy* (Open Court, 2000), The Simpsons *and Philosophy* (Open Court, 2001), The Matrix *and Philosophy* (Open Court, 2002), The Lord of the Rings *and Philosophy* (Open Court, 2003), and *The Philosophy of Martin Scorsese* (University Press of Kentucky, 2007).

R. Barton Palmer is Calhoun Lemon Professor of Literature at Clemson University in South Carolina. Among his many books on film are *Hollywood's Dark Cinema: The American Film Noir* (rev. ed., University of Illinois Press, forthcoming), *Joel and Ethan Coen* (University of Illinois Press, 2004), *Twentieth-Century American Fiction on Screen* (Cambridge University Press, 2007), and (with David Boyd) *After Hitchcock: Influence, Imitation, and Intertextuality* (University of Texas Press, 2006).

Steven M. Sanders is emeritus professor of philosophy at Bridgewater State College in Massachusetts. His work in ethics, political philosophy, and epistemology has appeared in numerous scholarly journals, including *Philosophia*, the *Journal of Social Philosophy,* and the *Southern Journal of Philosophy.* He has contributed essays to *The Philosophy of Film Noir* (2006), *The Philosophy of Neo-Noir* (2007), *The Philosophy of Stanley Kubrick* (2007), and *The Philosophy of Martin Scorsese* (2007), all published by the University Press of Kentucky. He is coeditor (with Aeon J. Skoble) of *The Philosophy of TV Noir* and (with R. Barton Palmer) of *Hitchcock as Moralist,* both forthcoming.

AEON J. SKOBLE is associate professor of philosophy and chair of the Philosophy Department at Bridgewater State College in Massachusetts. He is coeditor of the anthology *Political Philosophy: Essential Selections* (Prentice-Hall, 1999) and author of *Deleting the State: An Argument about Government* (Open Court, 2007), as well as many essays on moral and political philosophy in both scholarly and popular journals. In addition, he writes widely on the intersection of philosophy and popular culture, including such subjects as *Seinfeld, Forrest Gump, The Lord of the Rings,* superheroes, film noir, science fiction, and baseball, and he coedited and contributed to *Woody Allen and Philosophy* (Open Court, 2004) and the best-selling The Simpsons *and Philosophy* (Open Court, 2000).

KEVIN L. STOEHR is assistant professor of humanities at Boston University. He is the author of *Nihilism in Film and Television* (McFarland, 2006) and has written widely on popular culture and philosophy, including such topics as poker, Bob Dylan, *The Sopranos,* and Stanley Kubrick.

ALAN WOOLFOLK is professor of sociology and director of the core curriculum at Oglethorpe University in Atlanta. His most recent article is "The Horizon of Disenchantment: Film Noir, Camus, and the Vicissitudes of Descent," in *The Philosophy of Film Noir* (University Press of Kentucky, 2006). He is currently working on an introduction to the revised edition of R. Barton Palmer's *Hollywood's Dark Cinema: The American Film Noir* (University of Illinois Press, forthcoming) and editing the second volume of Philip Rieff's Sacred Order/Social Order series (University of Virginia Press, forthcoming). He is an advisory editor to *Society* and has twice been a National Endowment for the Humanities fellow.

INDEX

Abolition of Man, The, 179
A bout de souffle, 199
Adorno, Theodor, 3, 153, 157, 169
Against Interpretation, 199, 203
aggression, 9, 94–100, 160
A.I.: Artificial Intelligence, 165
Airstrip One, 178, 182, 187
Aldrich, Robert, 198
Alien, 11, 153
allegory, 4, 14, 59, 69, 92, 176, 212–13, 217
Allied Artists, 57
Alpha 60, 192, 195, 203
Alphas, 202
Alphaville, 192–200, 202–3
Alphaville, une etrange aventure de Lemmy Caution, 3, 12–14, 169, 191–96, 200–201, 203
alterity, 13, 171–72, 175, 188
Amis, Kingsley, 176
Anderson, Michael, 172, 177, 179–80
Animal Farm, 174, 179, 189n2
Anti-Sex League, 181
anxiety, 7, 73–81, 142
ape-men, 126, 162
a priori, 15, 214, 217–19
Archimedean, 10, 130
Arctic Ocean, 84
Aristotle, 48, 144
Army of the 12 Monkeys, 104, 107, 113–14, 116
artificial intelligence, vii, 1, 8, 10–11, 132–33, 135–37, 145, 149n16, 157, 165, 171, 193, 209

auteur, 196
authenticity, 48–49, 84
autonomy, 59, 93, 96

bad faith, 49
Bakhtin, Mikhail, 194
Baudelaire, Charles, 199
Becker, Ernest, 7, 76
Bedford Incident, The, 179
being-in-the-world, 48, 53, 74–75
being-toward-death, 74, 77–78, 84, 86
Bellamy, Edward, 178
Bergson, Henri, 195–96
big bang, 114
Biskind, Peter, 2, 59
Blade Runner, 2–3, 5, 12–13, 15, 18, 21–25, 27–35, 36n1, 36n4, 135, 153–54, 165, 169n3, 200
Bloom, Harold, 178
body criterion, 41, 42, 43
Bradbury, Ray, 115, 201
brain criterion, 42–43
branching universe model, 112–13, 116
Braucourt, Guy, 56
Brave New World, 173, 200–202
Bresson, Robert, 205n21
Breton, Andre, 200
Bukatman, Scott, 5
Build My Gallows High, 69
Burdick, Eugene, 179
Burgess, Anthony, 171, 175–76, 184, 189
Burnham, James, 173, 189n3

C-3PO, 153, 165
Cameron, James, 2, 17, 37n9, 99, 104,
    116n11, 135, 147n2, 169n2
Camus, Albert, 198, 201, 204, 225
Capitale de la douleur, 193
capitalism, 157, 172–73
Cartesian, 15, 50, 208, 216–17
categorical imperative, 169–70n11
causal loop, 114
Cavell, Stanley, 199
Christ, 4, 217–18
chronotope, 194–95
City of the Workers, 154, 164
Clavius, 126
Clockwork Orange, A, 153
cognition, 108, 130
Cold War, 4, 59, 69, 93, 95, 173,
    178–79, 184, 186, 189n2
Colossus: The Forbin Project, 10–11, 99
communism, 13, 59, 95, 175, 179
computer, 10–11, 14, 28, 99, 104,
    124–25, 129, 132–33, 192, 203,
    207, 209, 214–15, 218–29,
    221n16
Constantine, Eddie, 193, 198
containment, 91, 95
criteria of personhood, 39–41, 46–47,
    50, 52n1
cyberpunk, 22, 36n3
cyborg, 104, 156, 165

Dark City, 2–3, 5–6, 12, 21–24, 26–28,
    30–36, 36n1, 36n5, 153, 169n3,
    179, 182
Darke, Chris, 193
Dasein, 74–75, 129
Data, 165
Dawn of Man sequence (*2001: A Space
    Odyssey*), 122, 125, 128
Day the Earth Stood Still, The, vii, 3–4,
    8, 10–11, 91–92, 100, 100n3, 192

death, 4, 7–8, 28, 31, 64, 73–86, 108,
    129, 131, 197, 209
deconstruction, 51, 67
derangement, 67, 72n31, 191, 200
Derrida, Jacques, 6, 50–51
De Sade, Marquis, 159
Descartes, Rene, 14–15, 52, 129,
    207–8, 210, 214–19, 220n11
Dialectic of Enlightenment, 12, 153,
    157, 159, 161–63, 168–69
Dick, Philip K., 5
D.O.A., 72n31, 179
Dostoyevsky, Fyodor, 202
doublethink, 187
Dreyfus, Hubert, 130–33, 133n5
*Dr. Strangelove, or: How I Learned to
    Stop Worrying and Love the Bomb,*
    10, 153, 169n3
durée, 172, 195–96
dystopian fiction, 172, 175, 177, 188

Eagleton, Terry, 178
Ebert, Roger, 169n3
Einstein, 62
Eisenhower, Dwight D., 173
Ellison, Harlan, 147n2
Éluard, Paul, 191, 204
emotions, 5–6, 11, 27, 30, 33–35, 63,
    124, 133, 137–46, 148n11
empirical paradox, 9, 105, 107–9,
    115
Enlightenment, 12, 157, 160, 168, 172,
    177, 188
estrangement, 12, 83
Eternal Gardens, 164, 167
eternalism, 106–7
existentialism, 48, 51

facticity, 48–50, 52, 183
Fahrenheit 451, 201
Fail-Safe, 100, 101n12, 179

feminist, 7, 60, 63
film noir, 5, 7, 13, 22–25, 33, 36n4,
    55, 57, 64, 68–70, 71n1, 72n33,
    179–80, 186, 193–95, 198–99
Finney, Jack, 56, 115n2
Fordism, 173
Frankenstein, 1, 3–4, 7, 73–75, 77–85,
    165, 168–69
freedom, 49, 50, 95–100, 105, 109, 158,
    161, 178
Frye, Northrop, 172, 187

general relativity, 106
Gilmore, Richard A., 12, 72n30
Godard, Jean-Luc, 13, 192, 193, 196,
    200, 202–3
Grand Inquisitor, 202
grandfather paradox, 110–13
growing universe model, 111–12
Guy l'Éclair, 198

HAL, 124–25, 129, 132–33, 165
Hamilton, Linda, 136
harm principle, 94
Haut, Woody, 57
Heart Machine, 155
Heidegger, Martin, 7, 48, 73–77, 79,
    81, 84, 86, 129–30
Henderson, C. J., 16
heritage film, 13, 186–88
Higson, Andrew, 186, 188
Hobbes, Thomas, 96
Homer, 161
Horkheimer, Max, 3, 12, 153, 157–61,
    164, 168, 196
Human Jungle, The, 180
Hume, David, 46, 67
Hurt, John, 188
Huxley, Aldous, 173, 200

identity, personal, vii, 1, 3, 5–6, 17,

18n5, 24–25, 27, 30–32, 39–50, 59,
    64, 67, 105, 108–10, 115, 195
inauthenticity, 75, 84, 87n31
incoherence, 64–68
indiscernibility of identicals, principle
    of, 108
inference to the best explanation, 66
Ingsoc, 177, 180, 181, 183
Inner Party, 181, 182, 183,
innocent-on-the-run genre, 22, 24,
    36n5
Invasion of the Body Snatchers, vii, 2, 3,
    4, 7, 16, 55–70, 192
irony, 7, 10, 60, 69, 188

Jameson, Fredric, 184, 185, 199
Journey to the End of the Night, 200
Jurassic period, 103

Kaminsky, Stuart M., 56, 57
Kant, Immanuel, 52n1, 129, 159, 160,
    169n11
Karina, Anna, 200
Kauffmann, Stanley, 15
Kierkegaard, Søren, 49
Killers, The, 180
Kiss Me Deadly, 198
Klaatu, 4, 8–9, 91–100
Kubrick, Stanley, 2, 10, 15, 119–31,
    135

Labor Party, 186
Lang, Fritz, 1, 3, 157, 164, 165, 168
Lawrence, D. H., 176
Leibniz's law, 108, 109
Leninism, 173
Lewis, C. S., 179
Lewis, David, 108, 111
Locke, John, 32, 43
logical paradoxes, 105, 109, 113, 115
Lost Highway, 128

Lucas, George, 2, 15, 165
Lucky Jim, 176

Managerial Revolution, The, 173
Mainwaring, Daniel, 56, 57, 67, 69
Maltese Falcon, The, 198
Manichean, 198
Martin, Adrian, 200
Marx, Karl, 173
Maté, Rudolph, 179
Matrix, The, 2, 3, 14–15, 153, 207–10,
    214–19
Matrix Reloaded, The, 208
Matrix Revolutions, The, 208
McCarthy, Kevin, 16, 55, 57
memory criterion, 43–45
memory, 5, 6, 22–35, 43–45, 195, 201
Menippean satire, 172, 187
metaphysical paradox, 10, 105
metaphysics, vii, 14–15, 64, 111,
    207–10, 214, 216–17
Metropolis, 1, 3, 12, 15, 153–57, 162–68
Meursault, 201
Mills, C. Wright, 173
Miłosz, Czesław, 178
mind/body problem, 35, 220n11
Ministry of Love, 181, 182
modernity, 12, 14, 157, 159, 166, 182,
    183, 197
Mumford, Lewis, 174–75, 176
mythology, 12, 153, 158–62, 166–68,
    215

narrative, 7, 51–52, 56, 57, 120,
    122–25, 128, 158, 161, 171, 184,
    193–94
Neal, Patricia, 98
Newman, Kim, 4, 9, 10, 17
new wave, 196
Newtonian universe, 106
New Tower of Babel, 166–67

Newspeak, 181
Nexus 6 replicants, 24, 25, 28, 29
Nietzsche, Friedrich, 49, 131–32, 160,
    161, 198
1984 (film, 1956), 13, 177–83
1984 (film, 1984), 13, 183–89
1984 (novel), 3, 13, 172, 173–77
Niven, Larry, 17
North, Edmund H., 4
nostalgia, 174, 175, 177, 178, 184, 189

O'Brien, Edmond, 179, 180, 201
Oceania, 178, 180, 183, 184
off-world colony, 21, 27
One, the, 209–10, 218, 219
O'Neill, James, 4
Oracle, the, 209, 219, 220n7
Organization Man, The, 173
Orwell, George, 13, 172, 200
other bodies, problem of, 6
other minds, problem of, 6
Outer Party, 181, 182
Out of the Past, 69, 198

paradox of fiction, 3, 11, 136, 147
paranoia, 7, 50, 55, 67–70, 181, 186
Pascal's wager, 144
pastiche, 184, 188, 198–99, 200
Patalas, Enno, 163, 169n1
peace, 92–100
preemption, 9, 91, 97
personal identity, 1, 3, 5, 6–7, 27, 29,
    32, 39–52, 59, 64, 105, 107–15, 195
physical criteria, 41–43
Planet of the Apes, 192
Plato, 14–15, 52n1, 131, 210–14, 217
possibilism, 107
Pournelle, Jerry, 17
Power Elite, The, 173
praxis, 174
presentism, 107

progress, 14, 169n3, 172, 177, 183, 184, 188, 193, 195–97, 202
Promethianism, 76, 81
Proyas, Alex, 2, 21, 153
psychological continuity, 45–47
Pulp Fiction, 128

Radford, Michael, 13, 172, 183–89
Rascovich, Mark, 179
Reagan, Ronald, 9, 17
red scare, 4, 59
Rennie, Michael, 91
Ricoeur, Paul, 174–75
rights, 97–98, 101n6
Robocop, 165
robot, 8, 9, 91, 92, 95–99, 101n10, 133, 155, 157, 163
Rodden, John, 179
Rogin, Michael Paul, 59
Roswell, NM, 68

Santa Mira, CA, 16, 17, 55, 57, 58, 66, 68
Sartre, Jean-Paul, 48–49
Scalzi, John, 4
Schrader, Paul, 15
Schwartz, Delmore, 13
Schwarzenegger, Arnold, 40, 136
Scott, Ridley, 2, 5, 21, 33, 36n1, 135, 153, 200
Searchers, The, 10, 123, 128
Shelley, Mary, 65, 73, 101n11, 165
self-identity, 47–49, 51, 52
Shell Beach, 22, 29, 30, 32, 33, 35
Siegel, Don, 2, 55–57, 59, 60–62, 67, 69
Simon, John, 15
Singer, Irving, 10, 71n9
Sobchack, Vivian, 194
Solaris, 16
Sontag, Susan, 191, 192
Soviet Union, 17, 59, 95, 97, 99, 189n3

space-time, 106–7, 200
Spanish civil war, 94
special relativity, 106
Spicer, Andrew, 5
Spielberg, Steven, 15, 165
Star Trek: Generations, 165
Star Trek—The Motion Picture, 2
Star Wars, 2, 15, 153, 165
Steffen-Fluhr, Nancy, 61–63, 69–70
Steinhoff, William, 173, 189n3
Sterling, Jan, 180
Stranger, The, 201
Strategic Defense Initiative, 9, 17
subjectivity, 129–30

Tamiroff, Akim, 193
Taylor, Charles, 197
Taylor, Frederick, 173
tech noir, 136, 145–47
technology, 1, 8, 9, 10, 12, 14, 17, 73, 91, 92, 99, 100, 101n11, 120, 121, 125, 130–31, 153, 157, 158, 163, 165, 167, 203
Telotte, J. P., 176, 177
Terminator, The, vii, 2, 3, 10, 11, 29, 99, 104–15, 135–47, 165
Terminator 2: Judgment Day, 17, 116n11
Terminator 3: Rise of the Machines, 117n11
Things to Come, 192
Thomson, David, 15–16, 18n16
thrownness of being, 48
time machine, 103, 105, 106, 113
Time Machine, The, 178
time travel, 1, 3, 9–10, 103–17, 136, 147n3
Tolstoy, Leo, 145, 197
Total Recall, 2, 3, 6, 39–54
Tourneur, Jacques, 69, 198
12 Monkeys, 3, 9–10, 103–17

2001: A Space Odyssey, 2, 10–11, 15, 127–33, 165
transcendence, 81, 122, 132, 207, 217, 218

Verhoeven, Paul, 2, 40, 165

Wachowski, Andy, 2, 207, 208
Wachowski, Larry, 2, 207, 208
Walton, Kendall, 139, 148n4
Waltz of the Spaceships sequence (*2001: A Space Odyssey*), 126–27
Walzer, Michael, 178
Wanger, Walter, 56
Warburg, Fredric, 13

We, 200
weapons, nuclear, 9, 91, 98, 100
Weber, Max, 14, 196–98, 203
Wells, H. G., 178
Whale, James, 1, 73, 75–84
Wheeler, Harvey, 179
Williams, Bernard, 53n8
Williams, Raymond, 189n2
Wise, Robert, 2, 4, 91
Wittgenstein, Ludwig, 50
Wizard of Oz, The, 165
Wollen, Peter, 193
Wynter, Dana, 55

Zamyatin, Yevgeny, 200